# Art of Thinking

*Xing Zhou*

Math for Gifted Students

http://www.mathallstar.org

*use your mobile device to scan this QR code for more resources including books, practice problems, online courses, and blog.*

This book was produced using the LaTeX system.

# Contents

# Preface

Welcome to Math All Star© series!

Math All Star originates from a series of lectures given to a group of gifted middle school students with a love for mathematics and an interest in participating in competitions such as MathCounts, AMC, and AIME. These lectures aim to strengthen their problem-solving abilities and to introduce effective techniques that are not typically taught in the classroom.

As the popularity of Math All Star grew, the author began to upload lecture materials to create online courses, thereby providing students with the opportunity to progress at their own paces.

Since then, course materials have constantly been reviewed and updated to reflect student feedback and the observations made during lectures. Recent competition problems are also continuously analyzed and referenced to ensure the relevance of the contents. These course materials are the foundations of this Math All Star series.

Because competition math is a diversified subject that covers both a wide breadth and depth of topics, it is quite challenging to effectively cover all the material in one book that is appropriate for every interested student. Consequently, the author has decided to write a series of books, with each one focusing on a particular topic. Students are encouraged to pick and choose where to begin, depending on their individual skill levels and needs.

In addition to these books, the Math All Star website provides extra practice problems and serves as a highly recommended supplemental learning resource.

If there are any questions, comments, or concerns, please visit the website or email contact@mathallstar.org.

Happy learning!

*To visit the Math All Star website, scan this QR code or go directly to http://www.mathallstar.org*

# Chapter 1

# Introduction

Competition math is not about complicated theorems and formulas. Merely remembering hundreds, even thousands, of them is far from sufficient to win math competitions. Students must be able to think effectively in order to properly analyze and solve challenging math problems.

This book aims to help students understand some frequently used methodologies and improve their ability to think effectively. Contents in this book are organized based on methodologies rather than specific subjects. Consequently, examples and practice problems presented in each chapter may cover many different subjects. For example, *Chapter 8 Symmetry* contains problems relating to polynomial factorization, equation, counting, and so on. Despite falling into different areas, all of these problems can be solved by exploiting their intrinsic symmetric properties. Readers should focus on learning how to identify and utilize such properties. This skill is critical to develop in addition to specific math subject knowledge..

All the methodologies discussed in this book are intuitive and easy to understand. Some of them may be taught in regular classrooms. These include mathematical induction and proof by contradiction. Others are usually not taught in classrooms. Regardlessly, all of them are must-know for anyone who wants to become a strong contender in any math competition. In addition, mastering the art of thinking not only is helpful for improving students' contest performance during school years, but also has positive impacts on their future. For example, some problems in this book originate from various job interview questions. Being able to think effectively in order to solve such problems is certainly beneficial to their future.

# Chapter 2

# Go Simple and Go Extreme

## 2.1 The Golden Principle

Many competition problems are challenging. Some may even make students feel clueless and nowhere to start. Undoubtedly, a hard problem can only be solved after a student is able to make some breakthrough in understanding its nature. One effective approach to achieve this goal is to start considering the simplest and the most extreme scenario of the given problem.

> The golden principle of problem solving is to go simple and go extreme.

Let's illustrate this using the following example.

### Example 2.1.1

Joe invites you to play a game with him by placing quarters on a rectangular shaped table. Each person places one coin in turn. Coins cannot overlap. The one who cannot find enough space to place the next coin loses the game. Do you want to play first or let Joe to play first? What is your winning strategy?

This is neither a pure math problem, nor an extremely hard puzzle. However, it is an interesting example to study how the golden principle can help analyze and solve a problem in a systematic and effective manner.

Firstly, what is the simplest and the most extreme case in *Example 2.1.1*? It is when the table is merely big enough to hold one coin. Clearly, in such a case, the first player will win the game. Hence, an educated guess is that the person who plays first will be the winner.

Next, we need to validate our conjecture by investigating whether the first player can always win. If so, we need to find a winning strategy. If not, we need to figure out the condition under which the first player can win. Again, this goal can be achieved by looking at a few additional simple cases.

The $2^{nd}$ simplest cases is when the table can hold just two coins. In this case, if the first player places his coin at one end of this table (see the graph below, left), he is going to lose because his opponent can place a coin at the other end which will leave him no more place.

 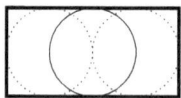

However, in this case, the first player can place his coin in the middle of the table which will secure his victory (see the graph above, right). Therefore, our conjecture that the first player wins still holds.

What if the table can hold three coins? Clearly, the first player still has a guaranteed win by placing his $1^{st}$ coin in the middle.

It appears that the first player has a winning strategy by placing his first coin at the center of the table. We note that the center is a special point on a rectangular-shaped table.

Let's investigate one more case and formalize our answer. When the table can hold four coins, placing the first coin at the center can still secure the victory for the first player regardless of this table's shape is $(2 \times 2)$ or $(4 \times 1)$. It now becomes apparent that after occupying the center spot, the first player can always find a vacant spot by following the symmetric rule if the second player can find his.

While obtaining the correct answer is important, an more important skill to learn is how to approach and solve such problems. As illustrated in this example, a good way to solve puzzling problems is to start analyzing simple and extreme situations. This will often lead to valuable insights and reveal important clues. For instance, in this example, studying the case of single-coin sized table immediately leads to the conjecture that the first player will win. Then, studying the possible winning strategy for the first player when the table can hold two coins reveals the importance of occupying the center. Further investigation of the cases of 3-coin and 4-coin sized tables helps validate and confirm the answer.

## 2.2   The Special Value Technique

A direct and powerful application of the go-simple-go-extreme principle is the *special value technique*. It is widely used to solve a variety of competition math problems. This method is based on the following conclusion.

> An assertion that holds in general cases must hold in special cases.

A typical use of the special value technique is to calculate the values of some to-be-determined constant parameters in a given relation. If such a relation is known to always hold, then by working on its simplest form may help determine those parameters in a less computationally intensive way.

Let's look at a couple of examples.

---

### Example 2.2.1

It can be shown that

$$1 \cdot 2^2 + 2 \cdot 3^2 + \cdots + n \cdot (n+1)^2 = \frac{n(n+1)}{12} \cdot (an^2 + bn + c)$$

holds for any positive integer $n$, where $a$, $b$, and $c$ are constants. Determine the values of $a$, $b$, and $c$.

---

It is possible to determine values of $a$, $b$, and $c$ by direct summarizing the left side and them comparing the corresponding coefficients[1]. However, given it is already known that this relation always holds, it will be easier to use the special value technique to find the answer.

---

[1]Methods to directly summarize such a sequence is discussed in the book *Power Calculation by Examples*.

Because three parameters need to be determined, there three special values are required. Obviously, 1, 2, and 3 are good candidates because they are the three smallest positive integers.

*Solution*

Setting $n = 1, 2$, and $3$, respectively:

$$\begin{cases} 1 \cdot 2^2 & = & \frac{1 \cdot 2}{12} \cdot (a + b + c) \\[2mm] 1 \cdot 2^2 + 2 \cdot 3^2 & = & \frac{2 \cdot 3}{12} \cdot (a \cdot 2^2 + b \cdot 2 + c) \\[2mm] 1 \cdot 2^2 + 2 \cdot 3^2 + c \cdot 4^2 & = & \frac{3 \cdot 4}{12} \cdot (a \cdot 3^2 + b \cdot 3 + c) \end{cases}$$

Solving this above system leads to $(a, b, c) = \boxed{(3, 11, 10)}$.

Done.

---

### Example 2.2.2

Compute $C_n^0 + C_n^1 + \cdots + C_n^n$.

---

It is obvious that some computation will be required if we simply replace $C_n^k$ with its definition as shown below:

$$C_n^0 + C_n^1 + \cdots + C_n^n = \frac{n!}{0!n!} + \frac{n!}{1!(n-1)!} + \cdots + \frac{n!}{n!0!} = \cdots$$

While looking for a simpler solution, we note that these $C_n^k$'s are coefficients in the following binomial expansion:

$$(1 + x)^n = C_n^0 + C_n^1 x + C_n^2 x^2 + \cdots + C_n^n x^n \qquad (2.1)$$

This is an identity which means it holds for any $x$. As a result, it will still hold if we assign a special value to $x$. A closer look at the right side of (2.1) will show setting $x$ to 1 can do the trick.

Setting $x = 1$ in *(2.1)* immediately leads to the answer:
$$C_n^0 + C_n^1 + \cdots + C_n^n = (1+1)^n = \boxed{2^n}$$

The $25^{th}$ problem in 2002 AMC 12A also offers a good example of applying the special value technique. This problem's main challenge is that both $P(x)$ and $Q(x)$ are unknown and abstract. This may make some students feel nowhere to start. However, setting $x = 1$ reveals a crucial relation between these two polynomials which can quickly lead to the answer.

## 2.3  Solve Functional Equation

A functional equation is an equation whose to-be-solved unknown is a function instead of a regular numerical variable. Some functional equations are relatively easy to solve. They can be tackled by treating the to-be-solved functions as regular variables and then applying usual equation solving techniques. Others can be very challenging to solve. In fact, solving functional equation frequently appear at national and international level math contests. For example, the $2^{nd}$ problem in 2017 IMO is a typical functional equation problem. Solving such problems usually relies on the special value technique.

Let's consider several examples at different challenging levels.

---

**Example 2.3.1**

Let function $f(x)$ satisfy the relation $f(x) + 2f(\frac{1}{x}) = 3$. Find the value of $f(2)$.

---

*Solution*

Letting $x = 2$ leads to
$$f(2) + 2f\left(\frac{1}{2}\right) = 3 \qquad\qquad (2.2)$$

Letting $x = \frac{1}{2}$ gives

$$f\left(\frac{1}{2}\right) + 2f(2) = 3 \tag{2.3}$$

Treating $f(2)$ and $f(\frac{1}{2})$ as two variables in the above two relations and solving them:

$$2 \times (2.3) - (2.3) \implies 3f(2) = 3 \implies f(2) = \boxed{1}$$

*Done.*

In the preceding example, setting $x = 2$ is a natural choice because the problem asks for $f(2)$. This leads to $f(\frac{1}{2})$ in addition to $f(2)$, hence we have to set $x = \frac{1}{2}$ to get another relation.

---

### Example 2.3.2

Let function $f(x)$ satisfy the relation $f(x) + 2f(\frac{1}{x}) = x$ for all non-zero real numbers $x$. Determine this function.

---

This example appears more abstract than the previous one. However, it can be solved by the same approach.

*Solution*

From the given condition, we have

$$f(x) + 2f\left(\frac{1}{x}\right) = x \tag{2.4}$$

and

$$f\left(\frac{1}{x}\right) + 2f(x) = \frac{1}{x} \tag{2.5}$$

(2.5) $\times 2-$ (2.4) yields

$$3f(x) = \frac{2}{x} - x \implies f(x) = \frac{2}{3x} - \frac{x}{3} = \boxed{\frac{2 - x^2}{3x}}$$

We can verify this function does satisfy the given condition:

$$f(x) + 2f\left(\frac{1}{x}\right) = \frac{2 - x^2}{3x} + 2 \times \frac{2 - \frac{1}{x^2}}{\frac{3}{x}} = \frac{2 - x^2}{3x} + \frac{4x^2 - 2}{3x} = x$$

*Done.*

These two problems are relatively simple. By setting $x$ to an appropriate value, additional relation can be derived from the original one which will result in a system of equations. Then it becomes possible to solve the target function using regular equation solving technique. The next example is more challenging. Carefully choosing appropriate special values is the key to solve such problems.

---

### Example 2.3.3

Find all functions $f(x)$ such that $f(0) = 1$, $f(\frac{\pi}{2}) = 2$, and for any real numbers $x$ and $y$,

$$f(x + y) + f(x - y) = 2f(x)\cos y$$

---

*Solution*

Setting $x = 0, y = \alpha$ where $\alpha$ is an arbitrary real number:

$$f(\alpha) + f(-\alpha) = 2f(0)\cos\alpha = 2\cos\alpha \qquad (2.6)$$

Setting $x = \frac{\pi}{2} + \alpha$ and $y = \frac{\pi}{2}$:

$$f(\pi + \alpha) + f(\alpha) = 0 \qquad (2.7)$$

Setting $x = \frac{\pi}{2}$ and $y = \frac{\pi}{2} + \alpha$:

$$f(\pi + \alpha) + f(-\alpha) = -2f\left(\frac{\pi}{2}\right)\sin\alpha = -4\sin\alpha \qquad (2.8)$$

Now, *(2.6) + (2.7) − (2.8)*:

$$2f(\alpha) = 2\cos\alpha + 4\sin\alpha \implies f(\alpha) = \cos\alpha + 2\sin\alpha$$

Replacing $\alpha$ with $x$ yields:

$$f(x) = \boxed{\cos x + 2\sin x}$$

Done.

General rules of choosing those special values are to utilize available information and simplify given equations. For instance, in the preceding example, because the values of $f(0)$ and $f(\frac{\pi}{2})$ are known, therefore both $x = 0$ and $x = \frac{\pi}{2}$ are natural choices. Meanwhiles, choosing $y = \frac{\pi}{2}$ makes sense because $\cos(\frac{\pi}{2}) = 0$.

## 2.4   Additional Application

Investigating extreme cases is not only useful to solve problems, but also helpful to understand intrinsic relationship among different concepts and theorems.

For example, in geometry, a tangent line of a circle can be viewed as an extreme case of a chord. To see this, let's start with a straight line intersecting a circle at two distinct points $A$ and $B$, and then move $B$ towards $A$. During this process, the chord $AB$ will get closer and closer to the tangent line passing $A$. When $B$ finally moves to the same spot as $A$ locates, the chord $AB$ becomes the tangent line $\ell$.

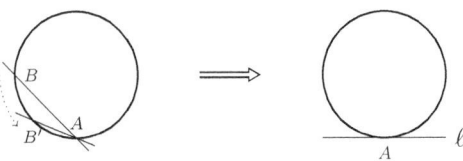

With this understanding, it will be easier to understand relationship among some tangent and chord related theorems. For instance, given two inscribed angles $\angle C$ and $\angle C'$ are known to be equal as shown below left, we can conjecture $\angle C = \angle ABB'$ below right by moving point $C'$ towards $B$.

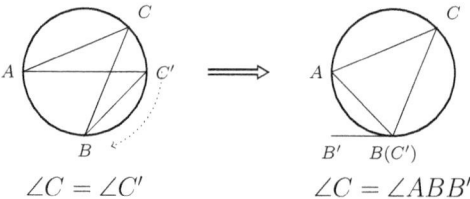

$$\angle C = \angle C' \qquad\qquad \angle C = \angle ABB'$$

Similarly, by treating a tangent line as an extreme case of a chord, the power of a point theorem can be unified without the need to distinguish the case of two chords and the case of one chord plus one tangent line.

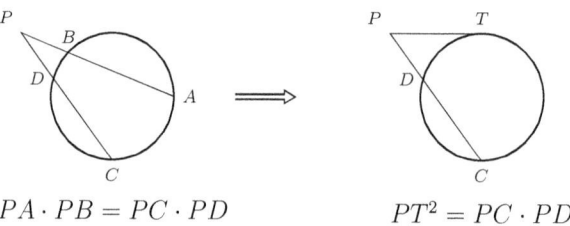

$$PA \cdot PB = PC \cdot PD \qquad\qquad PT^2 = PC \cdot PD$$

## 2.5 Practice

### Practice 1

Brahmagupta's formula states that the area of a cyclic quadrilateral whose sides lengths are $a$, $b$, $c$, and $d$ is given by

$$S = \sqrt{(p-a)(p-b)(p-c)(p-d)}$$

where $p = \frac{1}{2} \cdot (a+b+c+d)$.

Describe its relationship with the Heron's formula.

### Practice 2

It is known that the sum of the first $n$ cubes can be written as

$$1^3 + 2^3 + \cdots + n^3 = \left(\frac{1}{k} \cdot n \cdot (n+1)\right)^2$$

where $k$ is a positive constant. Determine the value of $k$.

### Practice 3

Evaluate the value of $C_n^0 + 2C_n^1 + 4C_n^2 + \cdots + 2^n C_n^n$.

### Practice 4

Find the sum of all the coefficients in the expanded form of

$$(x_1 + x_2 + \cdots + x_{2017})^{2017}$$

### Practice 5

Find the remainder when $x^{81} + x^{49} + x^{25} + x^9 + x$ is divided by $x^3 - x$.

### Practice 6

Find all functions $f : \mathbb{R} \to \mathbb{R}$ such that

$$x^2 f(x) + f(1 - x) = 2x - x^4 \qquad (2.9)$$

### Practice 7

Let the domain of function $f$ be all natural numbers and $f(1) = 1$. If for any natural numbers $m$ and $n$, it always hold that $f(m + n) = f(m) + f(n) + mn$, find $f(n)$.

### Practice 8

A chocolate bar is made up of a rectangular $m$ by $n$ grid of small squares. Two players take turns to break up the bar. In each turn, a player picks a rectangular piece of chocolate and breaks it into two smaller ones by snapping along one whole line of subdivisions between its squares. The player who makes the last break wins. Does one of the players have a winning strategy for this game?

## Practice 9

Suppose there are an infinite number of airports on a flat field which extends infinitely in all directions. No two of these airports are exactly the same distance apart. At a point in time, one plane will take off from each airport and land at its nearest neighboring airport. What is the maximum number of planes that may land at the same airport?

## Practice 10

There are 2017 people standing in a circle, numbered from 1 to 2017 sequentially. Starting from No. 1, they count alternating 1 and 2. People who count 2 will be out. The process continues until only one person left. What is his number?

## Practice 11

Let $p$ be an odd prime number. For positive integer $k$ satisfying $1 \leq k \leq (p-1)$, the number of divisors of $(kp+1)$ between $k$ and $p$, exclusive, is $a_k$. Find the value of $(a_1 + a_2 + \ldots + a_{p-1})$.

(2016 Japan MO)

## Practice 12

Find all functions $f : \mathbb{R} \to \mathbb{R}$ such that

$$f(yf(x) - x) = f(x)f(y) + 2x \qquad (2.10)$$

for all $x, y \in \mathbb{R}$.

(2016 Japan MO)

# Chapter 3

# Induction and Recursion

## 3.1 Base Case and Inductive Step

Mathematical induction is one of the most important methods of proof which is widely used in middle school, high school and beyond. Its core logic is relatively simple. There are two elements involved. One is a base case which can be verified by direct computation. The other is an inductive step which needs to be established. Then, by combining these two, an infinite recursion can be derived to show that an assertion is true for all positive integers.

> When a to-be-proved assertion is related to all positive integers, it is worth considering mathematical induction.

Let's first study an example.

---

**Example 3.1.1**

Prove
$$1 \cdot 1! + 2 \cdot 2! + \cdots + n \cdot n! = (n+1)! - 1 \qquad (3.1)$$

---

It is not obvious how to simplify the left side of *(3.1)* directly. However, by using the golden principle discussed in the previous chapter, we find it does hold for some simplest cases:

$$
\begin{aligned}
n = 1 &\implies 1 \cdot 1! & = 1 & = (1+1)! - 1 \\
n = 2 &\implies 1 \cdot 1! + 2 \cdot 2! & = 5 & = (2+1)! - 1 \\
n = 3 &\implies 1 \cdot 1! + 2 \cdot 2! + 3 \cdot 3! & = 23 & = (3+1)! - 1 \\
& \quad \cdots
\end{aligned}
$$

Based on these observations, it is likely that *(3.1)* is true for all positive integers. The question is how to prove this rigidly? Here comes the mathematical induction. It is a perfect choice in this case.

We have already verified that *(3.1)* holds when $n = 1$. If we can establish an induction to show *(3.1)* will hold for the next integer as long as it is true for the previous integer, then we can claim:

- Because it is true when $n = 1$, it will be true when $n = 2$,

- Then, because it is true when $n = 2$, it will be true when $n = 3$,

- Then, because it is true when $n = 3$, it will be true when $n = 4$,

- $\cdots$

This recursion can continue forever. Consequently, we can logically conclude that *(3.1)* holds for all positive integers.

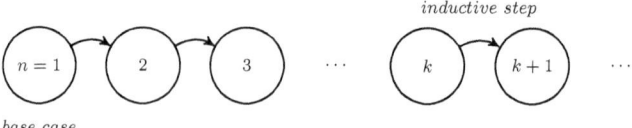

*base case*

Logistically, using mathematical induction requires three steps:

1) Establish a base case, i.e. the to-be-proved assertion is true when $n = 1$,

2) Assume the assertion is true when $n = k$,

3) Then show the assertion is true when $n = k + 1$

The reason that we can simply make the assumption without proving in the $2^{nd}$ step is because $k$ can initially equal 1 whose validity is already verified in the $1^{st}$ step. After the induction is derived in the $3^{rd}$ step, $k$ can be progressively substituted by 2, 3, and so on. Then, an infinite induction can be constructed to prove the validity of the assertion.

Let's illustrate this using a complete solution to *Example 3.1.1*.

*Proof*

Relation *(3.1)* holds when $n = 1$ because $1 \cdot 1! = (1 + 1)! - 1$.

Assume it holds when $n = k$, i.e.

$$1 \cdot 1! + 2 \cdot 2! + \cdots + k \cdot k! = (k + 1)! - 1$$

Then, when $n = k + 1$, we have

$$
\begin{aligned}
& 1 \cdot 1! + 2 \cdot 2! + \cdots + k \cdot k! + (k + 1) \cdot (k + 1)! \\
=& ((k + 1)! - 1) + (k + 1) \cdot (k + 1)! \\
=& (k + 1)! + (k + 1) \cdot (k + 1)! - 1 \\
=& (k + 2) \cdot (k + 1)! - 1 \\
=& (k + 2)! - 1 \\
=& ((k + 1) + 1)! - 1
\end{aligned}
$$

Hence, by the principle of mathematical induction, *(3.1)* holds for all positive integers.

*QED*

When mathematical induction is used, the target is already known. For instance, in the preceding example, the working objective when deriving the inductive step is to show $((k+1)+1)!-1$.

> The process of deriving the inductive step often involves a targeted transformation whose aim is to construct an already known result.

Let's review another example. In this example, because the target must contain $(k + 1)(k + 2)(2k + 3)$, we keep $(k + 1)$ as a common divisor and try to factorize the rest terms into $(k+2)$ and $(2k + 3)$.

---

### Example 3.1.2

Show that $1^2 + 2^2 + \cdots + n^2 = \frac{1}{6} \cdot n(n + 1)(2n + 1)$.

---

*Proof*

When $n = 1$, the left side equals 1 and the right side equals $\frac{1}{6} \times 1 \times 2 \times 3 = 1$. Hence the relation holds.

Assuming the relation holds when $n = k$, i.e.

$$1^2 + 2^2 + \cdots + k^2 = \frac{1}{6} \cdot k(k + 1)(2k + 1)$$

Then when $n = k + 1$, we have

$$
\begin{aligned}
&1^2 + 2^2 + \cdots + k^2 + (k + 1)^2 \\
=&\frac{1}{6} \cdot k \cdot (k + 1)(2k + 1) + (k + 1)^2 \\
=&\frac{1}{6} \cdot (k + 1)\Big(k(2k + 1) + 6(k + 1)\Big) \\
=&\frac{1}{6} \cdot (k + 1)(2k^2 + 7k + 6) \\
=&\frac{1}{6} \cdot (k + 1)(k + 2)(2k + 3)
\end{aligned}
$$

$$= \frac{1}{6} \cdot (k+1)((k+1)+1)(2(k+1)+1)$$

Therefore, by the principle of mathematical induction, this relation holds for all positive integer $n$.

QED

## 3.2  Variations

Mathematical induction is not only powerful, but also versatile. In addition to the basic form discussed in the previous section, there are some variations. For example,

  i) The base case does not have to be $n = 1$. For instance, to prove a claim such as "for any positive integer $n \geq 3, \cdots$ ", a base case $n = 3$ can be used.

 ii) There can be more than one base case, if needed. *Example 3.2.1* below uses this variation.

iii) The assumption does not need to be $n = k$. Instead, an assumption with respect to all $n \leq k$ can be used. This variation is utilized in *Example 3.2.2* below.

 iv) and so on

Let's now consider a couple of examples which utilize some of these variations.

---

### Example 3.2.1

Fibonacci sequence $\{F_n\}$ is defined as $F_1 = F_2 = 1$ and $F_n = F_{n-1} + F_{n-1}$ for $n > 2$. Show that

$$F_n = \frac{1}{\sqrt{5}}\left[\left(\frac{1+\sqrt{5}}{2}\right)^n - \left(\frac{1-\sqrt{5}}{2}\right)^n\right] \qquad (3.2)$$

---

Formula *(3.2.1)* can be directly derived using the technique discussed in the book *Competition Algebra*. Here, let's prove it using mathematical induction. Because $F_n$ depends on two previous terms, two base cases are required. Otherwise, with only one verified base case, the induction cannot be constructed.

*Proof*

The claim holds when $n = 1, 2$ because

$$\frac{1}{\sqrt{5}}\left[\left(\tfrac{1+\sqrt{5}}{2}\right)^1 - \left(\tfrac{1-\sqrt{5}}{2}\right)^1\right] = \frac{1}{\sqrt{5}} \cdot \sqrt{5} \qquad = 1$$

$$\frac{1}{\sqrt{5}}\left[\left(\tfrac{1+\sqrt{5}}{2}\right)^2 - \left(\tfrac{1-\sqrt{5}}{2}\right)^2\right] = \frac{1}{\sqrt{5}} \cdot \left[\tfrac{1+2\sqrt{5}+5}{4} - \tfrac{1-2\sqrt{5}+5}{4}\right] = 1$$

Assuming the claim holds when $n = k > 2$, i.e.,

$$F_k = \frac{1}{\sqrt{5}}\left[\left(\frac{1+\sqrt{5}}{2}\right)^k - \left(\frac{1-\sqrt{5}}{2}\right)^k\right]$$

then when $n = k + 1$,

$$F_{k+1} = F_k + F_{k-1}$$

$$= \frac{1}{\sqrt{5}}\left[\left(\frac{1+\sqrt{5}}{2}\right)^k - \left(\frac{1-\sqrt{5}}{2}\right)^k\right] + \frac{1}{\sqrt{5}}\left[\left(\frac{1+\sqrt{5}}{2}\right)^{k-1} - \left(\frac{1-\sqrt{5}}{2}\right)^{k-1}\right]$$

$$= \frac{1}{\sqrt{5}}\left[\left(\frac{1+\sqrt{5}}{2}\right)^{k-1}\left(\frac{1+\sqrt{5}}{2}+1\right) - \left(\frac{1-\sqrt{5}}{2}\right)^{k-1}\left(\frac{1-\sqrt{5}}{2}+1\right)\right]$$

$$= \frac{1}{\sqrt{5}}\left[\left(\frac{1+\sqrt{5}}{2}\right)^{k-1}\left(\frac{6+2\sqrt{5}}{4}\right) - \left(\frac{1-\sqrt{5}}{2}\right)^{k-1}\left(\frac{6-2\sqrt{5}}{4}\right)\right]$$

$$= \frac{1}{\sqrt{5}}\left[\left(\frac{1+\sqrt{5}}{2}\right)^{k-1}\left(\frac{1+\sqrt{5}}{2}\right)^2 - \left(\frac{1-\sqrt{5}}{2}\right)^{k-1}\left(\frac{1-\sqrt{5}}{2}\right)^2\right]$$

$$= \frac{1}{\sqrt{5}}\left[\left(\frac{1+\sqrt{5}}{2}\right)^{k+1} - \left(\frac{1-\sqrt{5}}{2}\right)^{k+1}\right]$$

Therefore, we conclude that *(3.2)* holds for all $n \geq 1$.

*QED*

The next example has two points worth noting. The first one is that the claim involves two variables. And, the second point is the proof uses the variation of $n < k$.

---

**Example 3.2.2**

Prove any positive proper fraction $\frac{m}{n}$ can be written as a sum of several reciprocals of distinct integers.

---

*Proof*

Let's apply induction on $m$ for any given $n$.

When $m = 1$, the claim obviously holds. Assume the claim holds for all integers less than $m$, we now show that it will hold for $m$ as well.

Writing $n$ as $n = qm - r$, where $q$ and $r$ are both integers, and $0 \leq r < m$.

If $r = 0$, then $\frac{m}{n} = \frac{m}{qm} = \frac{1}{q}$. The claim holds.

If $r > 0$, then

$$\frac{m}{n} = \frac{mq}{nq} = \frac{n+r}{nq} = \frac{1}{q} + \frac{r}{nq}$$

Because $r < m$, by the assumption, there exist several distinct positive integers, $1 < n_1 < n_2 < \cdots < n_k$ so that

$$\frac{r}{n} = \frac{1}{n_1} + \frac{1}{n_1} + \cdots + \frac{1}{n_k}$$

The reason that $n_1$ must be greater than 1 is because otherwise the sum will be no less than 1 which contradicts the fact $\frac{m}{n}$ is a proper fraction. It follows that

$$\frac{r}{qn} = \frac{1}{qn_1} + \frac{1}{qn_1} + \cdots + \frac{1}{qn_k}$$

All these numerators, $qn_i$'s, $(i = 1, 2, \ldots, k)$, are distinct and greater than $q$. Hence, we find $\frac{m}{n}$ can be expressed as the sum of the following:

$$\frac{m}{n} = \frac{1}{q} + \frac{r}{qn} = \frac{1}{q} + \frac{1}{qn_1} + \frac{1}{qn_1} + \cdots + \frac{1}{qn_k}$$

where all the numerators are distinct integers.

By the principle of mathematical induction, we conclude the assertion is true for any proper fraction number.

*QED*

## 3.3 Recursion

As a method of proof, mathematical induction requires the result to be known in advance. Therefore, it is usually not used to find the result. Notwithstanding that, the concept of induction can also be used to find the result.

For example, one can first make an educated guess of the result and then apply mathematical induction to prove his conjecture. Some practice problems in this chapter can be solved using this approach.

Additionally, it may be possible to establish and then solve the inductive relation to get the result. This approach is often called the recursion method. The next example is a typical problem that can be solved in this way.

---

### Example 3.3.1

Frank, the dog, can jump onto a staircase either one step a time or two steps a time. How many different possible ways for it to jump onto an 8-step staircase?

---

Let's first apply the go-simple-and-go-extreme principle to analyze this problem.

When there is only one step, obviously only one way exists. When there are two steps, there are two ways for Frank to jump onto this staircase: making two one-step jumps or making one two-step jump. When there are three steps, three possible ways can be found: 1-1-1, 1-2, and 2-1, where each figure indicates the number of steps Frank makes in that jump. When there are four steps, totally five possibilities exist: 1-1-1-1, 1-1-2, 1-2-1, 2-1-1, and 2-2.

It may be possible to continue this manual counting process. However, this approach is clearly not ideal when the number of steps increases to a large number.

One way to solve this problem neatly is to use the recursion method. In order to find this recursion relation, let's investigate how the result for a 3-step staircase can be obtained by earlier results. Because Frank can only jump either one or two steps a time, hence the last step before it jumps to the $3^{rd}$ step must be either the $1^{st}$ or the $2^{nd}$ step. It follows that the number of ways for Frank to land on the $3^{rd}$ step must equal the sum of the number of the ways for it to jump onto the $1^{st}$ and the $2^{rd}$ steps. Similarly, the total number of different ways to reach the top of a 4-step staircase equals the sum of different ways to jump onto a 2-step staircase and a 3-step staircase.

Here is the complete solution to this problem.

*Solution*

Let $F_k$ be the number of possible ways for Frank to climb a $k$-step staircase. Then, the target is to find the value of $F_8$.

It is clear that $F_1 = 1$ and $F_2 = 2$. When $k > 2$, the follow recursion should hold

$$F_k = F_{k-1} + F_{k-2} \tag{3.3}$$

This relation holds because the last step Frank must land on before it can jump onto the $k^{th}$ has to be either the $(k-1)^{th}$ or the $(k-2)^{th}$ step. There are $F_{k-1}$ and $F_{k-2}$ possibilities for Frank to reach these two steps, respectively. Therefore, there must be $(F_{k-1} + F_{k-2})$ ways to land on the $k^{th}$ step.

Based on *(3.3)*, the value of $F_8$ can be calculated recursively as shown below:

$$
\begin{aligned}
F_1 &= 1 \\
F_2 &= 2 \\
F_3 &= F_1 + F_2 &= 1 + 2 &= 3 \\
F_4 &= F_2 + F_3 &= 2 + 3 &= 5 \\
F_5 &= F_3 + F_4 &= 3 + 5 &= 8 \\
F_6 &= F_4 + F_5 &= 5 + 8 &= 13 \\
F_7 &= F_5 + F_6 &= 8 + 13 &= 21 \\
F_8 &= F_6 + F_7 &= 13 + 21 &= \boxed{34}
\end{aligned}
$$

*Done.*

The recursion method often relates to recursive sequence. For example, *(3.3)* and its two initial conditions, $F_1$ and $F_2$, in fact have defined a recursive sequence. It has the same recursion as Fibonacci sequence does, but with different initial values. Hence, they are two different sequences with different general formulas.

Techniques to solve recursive sequences has been discussed in the book *Competition Algebra*. However, in this particular example, the general formula to *(3.3)* can be obtained by that to the Fibonacci sequence, *(3.2)* on *page 21*. In order to achieve this, let's insert an artificial term in front of the solution to *Example 3.3.1*:

$$\boxed{1}, \quad \underbrace{1, \quad 2, \quad 3, \quad 5, \quad \cdots}_{original\ solution}$$

Now it is clear that the desired answer is the $9^{th}$ term in the Fibonacci sequence whose value can be computed using *(3.2)* directly.

$$F_n = \frac{1}{\sqrt{5}}\left[\left(\frac{1+\sqrt{5}}{2}\right)^n - \left(\frac{1-\sqrt{5}}{2}\right)^n\right]$$

When using *(3.2)*, we note that the odd terms in the $n-$power expansions will cancel each other and even terms will double. Therefore,

$$\begin{aligned}
F_9 &= \frac{1}{\sqrt{5}} \cdot \frac{2}{2^9}\left[C_9^1(\sqrt{5})^1 + C_9^3(\sqrt{5})^3 + C_9^5(\sqrt{5})^5 + C_9^7(\sqrt{5})^7 + C_9^9(\sqrt{5})^9\right] \\
&= \frac{1}{\sqrt{5}} \cdot \frac{1}{2^8} \cdot \sqrt{5} \cdot \left[9 + 84 \cdot 5 + 126 \cdot 25 + 36 \cdot 125 + 625\right] \\
&= \boxed{34}
\end{aligned}$$

This answer agrees with the previous result.

## 3.4   Practice

### Practice 1

Let $\{a_n\}$ be a sequence defined as $a_1 = 1$ and $a_n = \frac{a_{n-1}}{1+a_{n-1}}$ when $n \geq 2$. Find the general formula of $a_n$.

### Practice 2

An ATM machine can only dispense two-dollar bills and five-dollar bills. Show that it is always capable of dispensing exactly $n$ dollars when $n \geq 4$.

## Practice 3

Show that for any positive integer $n$, it is always hold that

$$2^n + 2 > n^2$$

## Practice 4

Show that it is always possible to cut $n$ squares of arbitrary sizes into some pieces and then use these pieces to construct a single bigger square.

## Practice 5

Show that

$$1 \cdot 2^2 + 2 \cdot 3^2 + \cdots + n \cdot (n+1)^2 = \frac{n(n+1)}{12} \cdot (an^2 + bn + c)$$

holds for any positive integer $n$, where $a$, $b$, and $c$ are constants.

## Practice 6

Find a prime number $p$ so that it always divides $(3^{2n+1} + 2^{n+2})$ where $n$ is a positive integer.

## Practice 7

Show that it always hold for any integer $n > 1$ that

$$\log_{10}(n!) > \frac{3n}{10} \cdot \left( \frac{1}{2} + \frac{1}{3} + \cdots + \frac{1}{n} \right)$$

## Practice 8

**(Tower of Hanoi)** Given a stack of $n$ disks of different sizes arranged in a neat stack in ascending order of size on one rod (the smallest at the top, thus making a conical shape), together with two empty rods, the towers of Hanoi puzzle asks for the minimum number of moves required to move the stack from one rod to another, where a move is only allowed if it places a smaller disk on a bigger one.

## Practice 9

Let $n$ be a positive integer. Show that

$$\left(1 + \frac{1}{3}\right)\left(1 + \frac{1}{3^2}\right) \cdots \left(1 + \frac{1}{3^n}\right) < 2$$

(China)

## Practice 10

Let $n$ be an integer greater than 2, prove $n^{n+1} > (n+1)^n$.

## Practice 11

Find all functions $f : \mathbb{Q} \to \mathbb{Q}$ such that the Cauchy equation

$$f(x + y) = f(x) + f(y)$$

holds for all $x, y \in \mathbb{Q}$ where $\mathbb{Q}$ is the set of all rational numbers.

# Chapter 4

# Proof by Contradiction

## 4.1 Attack from the Opposite

Some problems are difficult to tackle directly. One effective approach in such cases is to attack indirectly, i.e. from the opposite.

Let's take the following example to illustrate this approach.

---

**Example 4.1.1**

Show that $\sqrt{2}$ is irrational.

---

In addition to prove $\sqrt{2}$ is irrational directly, an alternative is to show it cannot be a rational number. This is because a real number must be either rational or irrational. If it cannot be rational, then logically we can assert it must be irrational.

*Proof*

If $\sqrt{2}$ is a rational number, then there must exist two relatively prime positive integers $m$ and $n$ so that $\sqrt{2} = \frac{m}{n}$. Squaring both sides and re-arranging yield $m^2 = 2n^2$. This implies $m$ is even. Let

$m = 2k$ where $k$ is an integer. Substituting $m$ with $2k$ leads to $4k^2 = 2n^2$, or equivalently, $2k^2 = n^2$. This means $n$ is even too.

However, two even numbers cannot be relatively prime to each other. This contradiction means the assumption that $\sqrt{2}$ is rational cannot be true. As a result, it must be true that $\sqrt{2}$ is irrational.

$$QED$$

The described approach is called proof by contradiction. Employing this method requires two steps:

1) Assume the opposite of the to-be-proved assertion holds,

2) Derive a contradiction based on the assumption made in the previous step

Then, we can claim the original assertion must be true.

## 4.2   More Examples

Proof by contradiction is an important method of proof. It can be used by its own or combined with other techniques. Let's review a couple of examples.

---

**Example 4.2.1**

Show that there exist infinite number of prime numbers.

---

*Proof*

If the claim is not true, then there exist only a limited number of primes. Let all these prime numbers be $p_1, p_2, \cdots, p_k$. Now, consider the following number

$$N = p_1 \cdot p_2 \cdots p_k + 1$$

Clearly, when $N$ is divided by any of these primes $p_i$, ($i = 1, 2, \cdots, k$), the remainder is always 1. This means that $N$ is not divisible by any of these known prime numbers. It follows that either $N$ itself is a prime or there is at least one more prime number in addition to all the previous listed $p_i$'s. Either way, it contradicts to the assumption that $p_1$, $p_2$, $\cdots$, $p_k$ are all the prime numbers. Hence, we conclude that the number of prime numbers is unlimited.

$$QED$$

In the next example, we first use the special value technique as discussed in *Chapter 2* to find some solutions to the given indeterminate equation, and then use the proof by contraction to show there is no other solution. Combining both, we can claim all solutions have been found.

---

### Example 4.2.2

Find all pairs of positive integers $(a, b)$ satisfying $a! + b! = a^b + b^a$.

---

*Solution*

If $a = b$, the equation reduces to $a! = a^a$. Because $a! < a^a$ holds for all $a \geq 2$, therefore there is only one solution in this case: $a = b = 1$.

When $a = 1$, the given equation reduces to $b! = b$ which gives an additional solution $(a, b) = (1, 2)$, By symmetry, $(a, b) = (2, 1)$ is a solution too.

Now let's prove these three pairs are only solutions using proof by contradiction. If there exists an additional solution, then $a$ and $b$ must not be equal and both of them are greater than 1. Without loss of generality, let's assume $1 < a < b$. It is easy to see $b!$ will be a multiple of $a$ as a result because

$$b! = 1 \cdot 2 \cdots a \cdot (a + 1) \cdots b \implies a \mid b!$$

Then, three terms in the relation $a! + b! = a^b + b^a$ are multiples of $a$. It must be true that the last term $b^a$ must be a multiple of $a$ too, i.e., $a \mid b^a$.

Let $p$ be any prime factor of $a$, then $p \mid b^a$ because $a$ is a multiple of $p$ and $b^a$ is a multiple of $a$. It follows that $p$ must divide $b$ due to the fact that $p$ is prime.

Next, we are going to examine the exponent of $p$ of both sides of the given equation.

$$a! + b! = a^b + b^a \implies a!\left(1 + \frac{b!}{a!}\right) = a^b + b^a \qquad (4.1)$$

On its left, we claim that $\left(1 + \frac{b!}{a!}\right)$ is not a multiple of $p$. This is because $p \mid b \implies p \mid \frac{b!}{a!} = (a+1) \cdots b$, hence $p$ cannot divide $\left(1 + \frac{b!}{a!}\right)$. Therefore, the exponent of $p$ in the left side's prime factorization equals that of $a!$'s prime factorization. It is well known by basic number theory that this exponent equals

$$\left\lfloor \frac{a}{p} \right\rfloor + \left\lfloor \frac{a}{p^2} \right\rfloor + \left\lfloor \frac{a}{p^3} \right\rfloor + \cdots \qquad (4.2)$$

where the flooring function $\lfloor x \rfloor$ returns the largest integer not exceeding $x$. It is clear that

$$\left\lfloor \frac{a}{p} \right\rfloor + \left\lfloor \frac{a}{p^2} \right\rfloor + \left\lfloor \frac{a}{p^3} \right\rfloor + \cdots < \frac{a}{p} + \frac{a}{p^2} + \frac{a}{p^3} + \cdots = \frac{a}{p-1} \le a$$

This means that the value of $(4.2)$ is less than $a$. However, on its right, because $a^b + b^a > a^a + a^a = 2a^a$, the exponent of $p$ in its prime factorization is at least $a$.

Hence, $(4.1)$ cannot hold. This contradiction means the assumption that more solution exists is false. It follows that the given equation has only three solutions $\boxed{(1,1), (1,2), (2,1)}$.

*Done.*

34

# 4.3 Practice

### Practice 1

There are 13 squares of side length 1 positioned inside a circle of radius 2. Show that at least two of these squares have a common point.

### Practice 2

Prove that no integers $x$ and $y$ can satisfy $x^2 - 4y = 2$.

### Practice 3

If there exist two integers $x$ and $y$ so that $ax + by = 1$ where both $a$ and $b$ are integers, show that $a$ and $b$ must be co-prime.

### Practice 4

If all sides of a convex pentagon $ABCDE$ are equal in length and $\angle A \geq \angle B \geq \angle C \geq \angle D \geq \angle E$, show that $ABCDE$ is a regular pentagon.

### Practice 5

Given $n > 2$ points on a plane. Prove if any straight line passing two of these points also passes another point, then all these $n$ points are collinear.

## Practice 6

Let the lengths of five line segments be $a_1$, $a_2$, $a_3$, $a_4$, and $a_5$, respectively, where $a_1 \geq a_2 \geq a_3 \geq a_4 \geq a_5$. If any three of these five line segments can form a triangle, then prove at least one of such triangles is acute.

## Practice 7

Prove if $x$ satisfies $0 < x < \frac{\pi}{2}$, then $\sin x + \cos x > 1$.

## Practice 8

Let $a_0$, $a_1$, $\cdots$, $a_n$ be all integers. If $a_0$, $a_n$ and the sum of $(a_0 + a_1 + \cdots + a_n)$ are all odd integers, show that the following equation has no rational roots:

$$a_0 x^n + a_1 x^{n-1} + \cdots + a_{n-1} x + a_n = 0$$

# Chapter 5

# Pigeonhole Principle

## 5.1  Pigeonhole Principle Explained

The pigeonhole principle is an intuitive and powerful tool to solve many interesting problems.

---

**Theorem 5.1.1   The Pigeonhole Principle**

If $(n+1)$ pigeons are placed in $n$ pigeonholes, then at least one pigeonhole will contain at least two pigeons.

---

When applying this principle to solve math problems, the pigeons and pigeonholes can be anything appropriate. For example, pigeons can be $n$ given numbers and pigeonholes can be a collection of distinct sets each of which contains all the numbers exhibiting certain properties. Let's consider an example.

---

**Example 5.1.1**

Show, that among any four integers, it is always possible to find at least two of them whose difference is a multiple of 3.

---

Clearly, the four integers are the pigeons. In order to use the pigeonhole principle, three pigeonholes are needed. Because there exist only three distinct possible remainders when any integer is divided by 3, we can construct three pigeonholes based on such remainders.

*Solution*

Construct three sets as below:

$$\{remainder = 0\}, \{remainder = 1\}, \{remainder = 2\}$$

Then any integer must belong to one of them according to the remainder when this integer is divided by 3. It follows that among any 4 integers, at lease two of them must belong to the same set, i.e. have the same remainders. The difference of these two numbers must be a multiple of 3.

*Done.*

> The key to employ the pigeonhole principle is to identify the pigeons and corresponding pigeonholes.

The pigeonhole principle has a couple of variations. They are both intuitive and self-explanatory.

### Theorem 5.1.2　The Pigeonhole Principle (Variation 1)

Let $m$ and $n$ be two positive integers. If $m \times n$ pigeons are placed in $n$ pigeonholes, then at least one pigeonhole will contain at least $m$ pigeons.

> **Theorem 5.1.3    The Pigeonhole Principle (Variation 2)**
>
> Let $m$ and $n$ be two positive integers. If $(m \times n + 1)$ pigeons are placed in $n$ pigeonholes, then at least one pigeonhole will contain at least $(m + 1)$ pigeons.

These two variations are also widely used.

## 5.2    More Examples

While pigeons are usually easy to identify, the pigeonholes may call for careful and creative construction. Let's review several examples to illustrate this.

---

### Example 5.2.1

Randomly select five points inside a unit square, show that at least two of these points are not separated by more than $\frac{\sqrt{2}}{2}$ in distance.

---

*Solution*

Divide this unit square into four smaller congruent squares as shown:

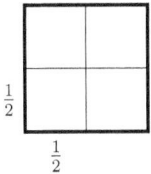

Let's treat these four smaller squares as pigeonholes and the five points as pigeons. Then, by the pigeonhole principle, at least two points will fall into the same smaller square (including its border). The maximum distance between two points in a square is the length of the square's diagonal. Hence, the distance between these two points can not be larger than $\frac{\sqrt{2}}{2}$.

*Done.*

---

### Example 5.2.2

Show that among any randomly selected 51 distinct numbers from $\{1, 2, \cdots, 100\}$, at least two of them are relatively prime.

---

*Solution*

Clearly, the 51 numbers are pigeons, we need to construct 50 pigeonholes each of which contains two relatively prime numbers. If so, by the pigeonhole principle, two of these numbers will be in the same "pigeonhole" and, therefore, are relatively prime.

Note that any two consecutive integers must be relatively prime. Therefore, the following 50 pairs are qualified pigeonholes:

$$\{1,2\}, \{3,4\}, \cdots, \{99, 100\}$$

*Done.*

---

### Example 5.2.3

Let $a_1$, $a_2$, $a_3$, $\cdots$, $a_9$ be a random permutation of 1, 2, 3, .., 9. Prove
$$(a_1 - 1)(a_2 - 2) \cdots (a_9 - 9)$$
is an even number.

---

*Solution*

All we need to show is that at least one of the $(a_k - k)$ is even where $k = 1, 2, \cdots, 9$.

Treat these 9 terms as 9 pigeonholes, we need to find 10 pigeons in order to apply the pigeonhole principle. Notice that there are five odd numbers in $1, 2, \cdots, 9$. Therefore, there are five odd numbers in $a_1, a_2, \cdots, a_9$, too. These 10 odd numbers can be used as our pigeons.

By the pigeonhole principle, at least two of these odd numbers will be grouped in one bracket. As a result, their difference must be even. This proves the claim.

*Done.*

---

**Example 5.2.4**

Prove that among any randomly selected 51 numbers from 1, 2, 3, ..., 100, there must exist two numbers among which one divides the other.

---

*Solution*

Any positive integer can be expressed in the form of $j \cdot 2^k$ where integer $j$ is odd and integer $k$ is non-negative. Hence, we create the following 50 sets to be used as pigeonholes:

- $1 \times 2^k : \{1, 2, 4, 8, 16, 32, 64\}$

- $3 \times 2^k : \{3, 6, 24, 48, 96\}$

- $5 \times 2^k : \{5, 10, 20, 40, 80\}$

- $\cdots$

- $99 \times 2^k : \{99\}$

The union of these 50 sets includes all the integers between 1 and 100. Hence, by the pigeonhole principle, at least two of these 51 numbers will fall into the same set. Clearly, by the construction of these sets, the bigger number must be a multiple of the smaller one.

*Done.*

---

### Example 5.2.5

Let $\{a_1, a_2, \cdots, a_7\}$ be a sequence of any 7 randomly chosen integers. Show that there always exist one or more consecutive numbers in this sequence whose sum is a multiple of 7.

---

*Solution*

Consider the following 7 sums:

- $S_1 = a_1$

- $S_2 = a_1 + a_2$

- $\cdots$

- $S_7 = a_1 + a_2 + \cdots a_7$

If any of these 7 sums is a multiple of 7, then the claim holds. Otherwise, the remainders of these sums divided by 7 can only be 1, 2, $\cdots$, or 6. Therefore, at least two of them will have the same remainder which implies their difference is a multiple of 7.

Meanwhile, by the construction of these 7 sums, it is obvious that the difference of any two such numbers must be a sum of one or more consecutive elements in the original sequence.

*Done.*

The conclusion of this example can be generalized as: for an $n$-element integer sequence, there must exist one or more consecutive elements whose sum is a multiple of $n$.

# 5.3 Practice

### Practice 1

Show that among any randomly placed 5 points inside an equilateral triangle whose side length is 2, the shortest distance between any two of these five points can not be longer than 1.

### Practice 2

Nine points are randomly placed in a unit square. Show that it is possible to select three of them to form a triangle whose area is no larger than $\frac{1}{8}$.

### Practice 3

There are only two problems in a math test. Ten points will be awarded for every correct answer. Two points will be given for any skipped problem. No point will be given for wrong answer. The teacher claims there must be at least 3 students who will receive a same score. Can you figure out the minimal number of students taking this test?

### Practice 4

A box contains a large quantity of four different types of Easter eggs. One kid is allowed to take one or two eggs of his choice from this box. What is the minimal number of kids must be there in order to confidently assert that at least two kids make the same choice?

### Practice 5

Given any five integers, show that it is always possible to select three of them so that their sum is a multiple of 3.

### Practice 6

Show that among any $(n + 1)$ integers, one can always find two of them whose difference is divisible by $n$.

### Practice 7

Given 12 different 2-digit integers, show that one can always choose two of them so that their difference is a two-digit integer with identical unit and tens digits.

### Practice 8

Show that there exists a multiple of 2017 whose digits are either 8 or 0.

## Practice 9

Show that any convex pentagon must have three vertices $A$, $B$, and $C$ such that $\angle ABC \leq 36°$.

## Practice 10

Each point of a circle is colored either red or blue.

(a) Prove that there always exists an isosceles triangle inscribed in this circle such that all its vertices are colored the same.

(b) Does there always exist an equilateral triangle inscribed in this circle whose three vertices are colored the same?

(Philippines)

## Practice 11

A chess board is an $8 \times 8$ grid. A bishop can attack another piece on the same diagonal as it locates. What is the maximum number of bishops can be placed on a chess board peacefully with no one being able to attack another?

## Practice 12

Show that in a $n$-people party, at least two of them have met the same number of other guests before.

## Practice 13

Two different sized roulettes share the same center. Each is divided into 200 equal sections. On the larger roulette, 100 randomly picked sections are colored in red and the rest are colored in green. The 200 sections on the smaller roulette are colored randomly in either red or green. The number of red sections and that of green section may be different. Show that it is always possible to rotate these two roulettes to a position so that at least 100 paired sections have the same color.

## Practice 14

Show that no matter how 10 people form a queue, there must exist at least four of them whose heights are in either ascending or descending order. These four people may or may not be next to each other.

## Practice 15

Show that every integer $k > 1$ has a multiple which is less than $k^4$ and can be written in base 10 using at most 4 different digits.
(IMO Short List Problem)

# Chapter 6

# The Coloring Method

## 6.1 Examples

The coloring method is a visualized approach to solve some math problems and brain teasers. It is able to discover some intrinsic properties by coloring a graph into certain patterns. By doing so, the given problem can be solved in an obvious and intuitive way.

> One picture is worth a thousand words.

Let's consider several examples.

### Example 6.1.1

Joe cuts the top left corner and the bottom right corner of an $8 \times 8$ board. He then tries to cover the remaining using thirty-one $1 \times 2$ pieces. Is it possible? Note: a $1 \times 2$ piece can be rotated, but cannot be further broken up.

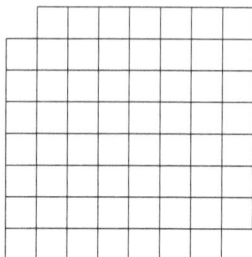

The answer is no. To prove this conclusion, let's color this board in alternating black and white just like a chess board.

Obviously, there are 32 black squares and 30 white ones. However, a $1 \times 2$ piece covers 1 black and 1 white square. This means that 31 such pieces will cover 31 squares of each color. Hence, it is clear that it is an impossible task to cover this board using thirty one $1 \times 2$ pieces.

> The key to solve such problems is to design an appropriate coloring pattern.

The following is a similar problem which requires a different coloring scheme.

**Example 6.1.2**

This time, Joe cuts a $2 \times 2$ corner off an $8 \times 8$ board, and then tries to cover the remaining board using 15 L-shaped pieces made of 4 grids as shown. Is it possible?

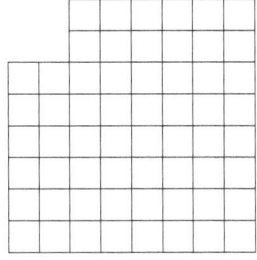

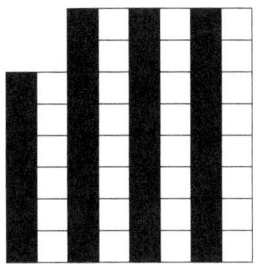

This is also an impossible task. In order to prove this, let's color this board in the following manner.

It is clear that there are equal numbers of black squares and white ones this time. Meanwhile, regardless how a L-shaped piece is placed, it will always cover 3 squares of one color and 1 square of the other, creating a difference of 2. Because 15 is an odd number, it is impossible to get a sum of 0 by adding fifteen $\pm 2$. This means it is impossible to use 15 such pieces to cover equal numbers of black and white squares. Hence, the conclusion holds.

Another popular technique to employ the coloring method is to use it with the Pigeonhole principle. The next example is a typical problem which can be solved in this way.

**Example 6.1.3**

Show that among any 6 people in the world, there must exist 3 people who either know each other or do not know each other.

(Hungarian)

Let's use six dots to represent any six people. If two people know each other, we connect these two dots using a dashed line. Otherwise, if they do not know each other, these two dots are connected using a sold line.

By using this convention, the original problem is equivalent to showing that there must exist a triangle whose three sides are either all solid or all dashed.

Let's consider person $A$. There are five connections from $A$ to the other five people each of which is either solid or dashed. By the pigeonhole principle, at least three of such connections must be the same type. Without loss of generality, let's assume $A$ is connected with $B$, $C$, and $D$ by solid lines.

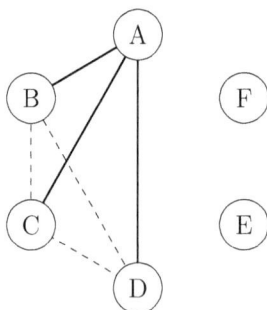

Now consider the three connections among $B$, $C$, and $D$. If any of them is solid, then this connection's two ending points and point $A$ form a triangle with all solid sides. Otherwise, if none of them is solid, then $B$, $C$, and $D$ form a triangle whose sides are all dashed.

## 6.2 Practice

### Practice 1

There are 24 lily pads on a pond as shown below. Frank, the frog, wants to visit all these pads without stopping at the same pad more than once. He can jump to a neighboring pad either horizontally or vertically, but not diagonally. If Frank can choose any pad as its starting point, can he achieve his goal?

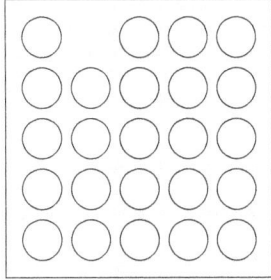

### Practice 2

In the center unit of a $3 \times 3 \times 3$ cubic lives a bug. Two units which share a face are connected via a door. The bug wants to visit all the units starting from his own without repeating. Is it possible?

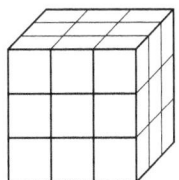

### Practice 3

Show that if an $m \times n$ grid can be completely covered by some L-shaped grids consist of 4 unit grids without overlapping, then the product of $m$ and $n$ must be a multiple of 8.

(Beijing)

### Practice 4

Show that it is impossible to cover an $8 \times 8$ board using fifteen $4 \times 1$ pieces and one $2 \times 2$ piece.

### Practice 5

Is it possible to cover a $6 \times 6$ grid using one L-shaped piece made of 3 grids and eleven $3 \times 1$ smaller grids?

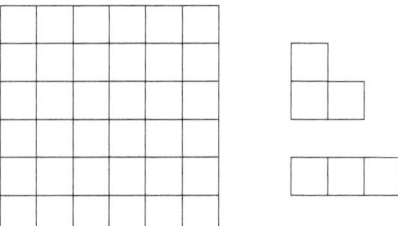

### Practice 6

It is possible to use some $1 \times 2 \times 4$ blocks to construct a $6 \times 6 \times 6$ cubic?

## Practice 7

There are 6 points in a 3-D space. No three points are on the same line and no four points are one the same plane. Hence totally 15 segments can be created among these points. Show that if each of these 15 segments is colored either black or white, there must exist a triangle whose sides are of same color.

## Practice 8

Seventeen people correspond by mail with one another - each one with all the rest. In their letters only three different topics are discussed. Each pair of correspondents deals with only one of these topics. Prove that there are at least three people who write to each other about the same topic.

(1964 IMO)

## Practice 9

Randomly color all the points in a plane using either white or black. Show that

1. There must exist a unit length segment whose two ends are colored same.

2. There must exist a right triangle whose three vertices are colored same.

## Practice 10

Is it possible to arrange the numbers $1, 1, 2, 2, 3, 3, \cdots,$ 1986, 1986 into a sequence so that there is 1 number between two 1's, 2 numbers between two 2's, $\cdots$, 1986 numbers between two 1986's?

(China)

# Chapter 7

# Two-State Problem

## 7.1 Introduction

Many objects have properties which can have only two possible values or states. For example, an integer can be either even or odd. A light can be either on or off. Tossing a coin will result in either heads or tails. This fact can be used to create many interesting competition problems and brain teasers which are called "two-state" problems. Quite often, solving such two-state problems requires no complex mathematics theorems and formulas, but creative thinking.

This chapter focuses on two topics of this subject. The first is to solve odd-even parity related problems. The second is to build a mathematical model to tackle some brain teasers.

## 7.2 Odd Even Parity

An integer can be either even or odd. It is easy to derive the following conclusions where $E$ indicates an even integer and $O$ indicates an odd one:

- Addition / Subtraction: $E \pm E = E, O \pm O = E, O \pm E = O$

- Multiplication; $E \times E = E, O \times O = O, E \times O = E$

By extension,

i) If the sum of two integers is even, then the two integers must have the same parity.

ii) If the sum of two integers is odd, then one of the them must be even and the other must be odd.

iii) The product of several integers is odd if all of them are odd.

iv) The product of several integers is even if and only if at least one of them is even

v) Adding an integer with an even integer does not change its parity

vi) Multiplying an integer with another odd integer does not change its parity.

vii) An integer $N$ always has the same parity as its opposite number $-N$.

The last property can be derived in many ways. For example $-N = N - (2N)$ where $2N$ must be even. Hence, these two numbers must have same parity. Alternatively, $-N = N \times (-1)$ where $-1$ is odd. Hence, we can conclude $-N$ must keep the same parity as $N$.

Let's consider a couple of examples.

**Example 7.2.1**

Randomly putting either '+' or '-' in front of each of 1, 2, $\cdots$, 2017, and then add these 2017 modified numbers together. Which of the following statements is correct:

1) The sum is even

2) The sum is odd

3) The sum can be either even or odd

*Solution*

Because $-N$ has the same parity as $N$, therefore the sum will have the same parity as

$$1 + 2 + \cdots + 2017 = \frac{(1 + 2017) \times 2017}{2} = 1009 \times 2017$$

which is odd.

Therefore the correct answer is (2).

*Done.*

The sum of $(1+2+\cdots+2017)$ is odd can also be shown by the fact that it is a sum of 1009 odd numbers and 1008 even ones. Adding an odd number of odd integers results an odd sum. Meanwhile, the sum of any number of even numbers is even.

**Example 7.2.2**

Let $a$ and $b$ be two positive integers and satisfy

$$123456789 = (11111 + a)(11111 - b)$$

Show that $(a - b)$ is a multiple of 4.

(Japan)

*Solution*

From the given condition, we have

$$11111 \times (a - b) = ab + (123456789 - 11111 \times 11111)$$
$$11111 \times (a - b) = ab + 4 \times 617$$

Now we claim that $(a - b)$ must be even. If this is not true, then the left side is odd. It follows that $ab$ on the right side must be odd which implies both $a$ and $b$ are odd. However, this will make $(a - b)$ even which contradicts the assumption. Hence, $(a - b)$ must be even.

When $(a - b)$ is even, the left side is even. This implies $ab$ is even, or at least one of them is even. If one of them is even, and their difference $(a - b)$ is also even, then both of them must be even.

If both $a$ and $b$ are even, then $ab$ must be a multiple of 4. Hence, the right side is a multiple of 4. This will lead to the conclusion that $(a - b)$ on the left is a multiple of 4 because 11111 is not divisible by 2.

*Done.*

## 7.3 The $\pm 1$ Model

It may be easy to find the answer to some two-state problems by educated guess. However, presenting convincing reasoning in a clear and simple way may sometimes be challenging. In such cases, it may be convenient to employ the $\pm 1$ model which involves mathematic computations using $\pm 1$.

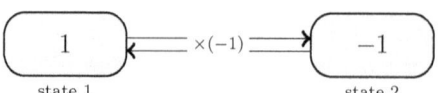

This technique is illustrated in the next example.

---

**Example 7.3.1**

There are nine coins on the table. All are heads up. In each move, you can flip two coins. What is the minimal number of moves to make them all heads down?

---

After a few tries, it is not difficult to make an educated guess that it is impossible to turn all the nine coins heads down. The question is how to prove this conjuncture clearly and simply?

*Solution*

There are two states of a coin: either heads up or heads down. Let's use $(1)$ to indicate heads up and $(-1)$ to indicate heads down. Each flip changes $(1)$ to $(-1)$, or vice versa. Therefore, a flip can be modeled as multiplying by $(-1)$.

Under this model, the initial values of all the nine coins are $(1)$ because they are all heads up. The product of all these nine values is $(1)$. If they all become heads down, then all their values will be $(-1)$ which makes their product $(-1)$. However, in each move, two flips are performed. This is equivalent to making two "multiplying by $(-1)$" operations. Multiplying two $\times (-1)$ equals multiplying $(1)$. Hence, it will not change the overall product of these nine values. The means the product of all the nine values will always be $(1)$. Therefore, it is impossible to reach the desired end state, i.e. all nine coins are heads down.

*Done.*

## 7.4 Invariant

Invariant means that something never changes. In the context of math, invariant means a function, quantity, or property that remains unchanged when a specific transformation is applied.

Identifying invariant often plays a critical role in solving some problems. For example, in *Example 7.2.1* on *page 57*, the parity of the sum is an invariant regardless of the sign put in front of each number. In *Example 7.3.1* on *page 59*, the product of all the individual states is an invariant.

---

### Example 7.4.1

Start with the numbers $1, 2, 3, \cdots, 2017$ and perform the following operation: cross any two randomly selected numbers and replace with the absolute value of their difference. Do this until there is only one number reamins. What is the parity of the last number?

---

Similar to *Example 7.2.1* on *page 57*, the parity of the sum is an invariant here too.

*Solution*

First, we claim that the parity of $|x - y|$ is the same as $(x - y)$. To prove this claim, we note that $|x - y|$ equals either $(x - y)$ or $(y - x)$. These two values are opposite numbers, thus have the same parity. This leads to the conclusion that $|x - y|$ has the same parity as $(x - y)$.

Then, we claim that $|x - y|$ has the same parity as $(x + y)$ because $(x + y)$ has the same parity as $(x - y)$. This reasoning holds because the difference between $(x + y)$ and $(x - y)$ is an even number.

Now, it follows that replacing any two integers $x$ and $y$ with $|x - y|$ will persevere the parity of their sum. Therefore, the parity of the last remaining number is the same as $(1 + 2 + \cdots + 2017)$

which is odd.

*Done.*

## 7.5  Practice

**Practice 1**

Find all pairs of prime numbers $(x, y)$ so that $x^2 + y = 2003$.

**Practice 2**

Four $x$'s and five $o$'s are written around the circle in an arbitrary order. If two consecutive symbols are the same, then a new $x$ is inserted in between. Otherwise, a new $o$ is inserted. After nine new symbols are inserted, the previous 9 old ones are erased. Is it possible to get nine $o$'s after having repeated this operation for a finite time?

**Practice 3**

There are three piles of stones, numbering 19, 8, and 9, respectively. You are allowed to choose two piles and transfer one stone from each of them to the third pile. Is it possible to make all piles all contain exactly 12 stones after several such operations?

**Practice 4**

Prove: if $a$, $b$, $c$ are odd integers, then the quadratic equation $ax^2 + bx + c = 0$ has no integer solution.

### Practice 5

Let $f(x)$ be a polynomial with integer coefficients. If $f(1) = 3$, is it possible that $f(3) = 0$?

### Practice 6

If the sum of $n$ integers equals 0 and their product equals $n$, show that $n$ must be a multiple of 4.

### Practice 7

Given any nine distinct points in 3-dimensional space with integer coordinates, show that there must exist two points so that the segment between them contains another point of integer coordinates.

### Practice 8

There are four types of cards costing 10, 15, 25 and 40 cents each. Joe bought totally 30 cards using several dollar bills exactly. If he bought 5 each of two types, and 10 each of the other two types, how much total did he spend on this purchase?

### Practice 9

There are $n$ points, $A_1, A_2, \cdots, A_n$ on a line segment, $\overline{A_0 A_{n+1}}$. The point $A_0$ is black, $A_{n+1}$ is white, and the rest points are colored randomly either black or white. Show that, among these $n+1$ line segments $A_k A_{k+1}$, where $k = 0, 1, \cdots, n$, the number of segments with different colored ending points is odd.

## Practice 10

Given two grids shown below, is it possible to transform (a) to (b) after a series of operations? In each operation, one can change all the signs in either one entire row or one entire column.

(Russia)

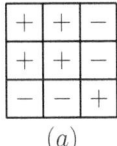
(a)

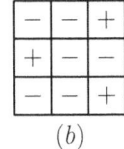
(b)

## Practice 11

Let $0.a_1a_2a_3a_4 \cdots$ be a decimal where $a_1$ is an odd integer and $a_2$ is an even integer. For $k > 2$, $a_k$ equals the unit digit of the value $(a_{k-1} + a_{k-2})$. Prove such a decimal must be rational.

## Practice 12

Katie had a collection of red, green and blue beads. She noticed that the number of beads of each color was a prime number and that the numbers were all different. She also observed that if she multiplied the number of red beads by the total number of red and green beads she obtained a number exactly 120 greater than the number of blue beads. How many beads of each color did she have?

(Scottish)

## Practice 13

An executioner lines up 100 prisoners single file and puts a red or a blue hat on each prisoner's head. Every prisoner can see the hats of the people in front of him in the line - but not his own hat, nor those of anyone behind him. The executioner starts at the end (back) and asks the last prisoner the color of his hat. He must answer "red" or "blue." If he answers correctly, he is allowed to live. If he gives the wrong answer, he is killed instantly and silently. (While everyone hears the answer, no one knows whether an answer was right.) On the night before the line-up, the prisoners confer on strategy to help them. What should they do in order to save as many prisoners as possible?

# Chapter 8

# Symmetry

## 8.1 Beyond Geometric Symmetry

The concept of symmetry originates from geometry. It is not surprising that many geometry problems involve symmetric shapes. As a result, exploiting symmetry to solve geometry problems is a big topic and is discussed in the book *Geometry Techniques* written by the same author. The focus here is to extend the concept of symmetry beyond geometry in order to solve other types of problems.

Generally speaking, a problem can be called symmetric if some of its parts can be switched without changing the nature of this problem. For instance, the following is a symmetric polynomial because switching $a$ and $b$ will result in an identical polynomial

$$(a + b)^2 = a^2 + 2ab + b^2$$

It is worth pointing out that symmetry is not limited to just two parts. For example, the following is a symmetric equation with three variables:

$$\frac{1}{x} + \frac{1}{y} + \frac{1}{z} = \frac{3}{5}$$

Discovering and utilizing a given problem's intrinsic symmetric properties are often helpful to solve some competition problems.

## 8.2 Solving Algebra Problems

### 8.2.1 Symmetric Polynomial Factorization

Factorizing a polynomial can be challenging. However, mastering polynomial factorization is important for anyone who wants to perform well in math competitions. There are many different polynomial factorization techniques. This section discusses how to effectively factorize symmetric polynomials.

It is obvious that all divisors of a symmetric polynomial must be symmetric. This is because if switching two variables does not change a polynomial's original form, then doing so must not alter the factorized form either. Meanwhile, when the number of variables and the degree of the polynomial are given, there only exists a limited number of possible symmetric factorization forms.

Based these two above insights, it may be possible to make an educated guess first when attempting to factorize a symmetric polynomial. Such approach is often proved to be effective in practice.

Let's illustrate this technique using the following example.

---

**Example 8.2.1**

Factorize $(ab + bc + ca)(a + b + c) - abc + (a + b)(b + c)(c + a)$.

---

For convenient, let's use $f(a, b, c)$ to denote this $3^{rd}$ degree polynomial. If it can be factorized, then it can only be either a product of three $1^{st}$ degree polynomials or a product of one $1^{st}$ degree and one $2^{nd}$ degree polynomial. Therefore, it can only be factorized into one of the following three forms:

1. $k(a + b + c)^3$

2. $k(a + b)(b + c)(c + a)$

3. $k(a + b + c)(a^2 + b^2 + c^2 + m(ab + bc + ca))$

where $k$ and $m$ are to-be-determined constants.

If it is the $1^{st}$ case, then $(a+b+c)$ divides $f(a, b, c)$. This means when $(a + b + c) = 0$, $f(a, b, c)$ must always equal 0. However, this does not hold because in such situation, we find

$$f(a, b, c) = 0 - abc + (-c)(-a)(-b) = -2abc$$

which is not necessarily equals 0. Thus, the $1^{st}$ form is not correct. By the same reason, the $3^{rd}$ form can be ruled out too.

This leaves the $2^{nd}$ to be the only possibility. To be sure, let's check whether $(a + b)$ can divides $f(a, b, c)$. If it does, then $(b + c)$ and $(c + a)$ must divide $f(a, b, c)$ too because of symmetry. If all of them divide $f(a, b, c)$, then $f(a, b, c)$ must be a multiple of $(a + b)(b + c)(c + a)$. Because both $f(a, b, c)$ and $(a + b)(b + c)(c + a)$ are $3^{rd}$ degree, their difference can only be a numerical constant.

It is indeed this case because when $(a + b) = 0$, we find

$$f(a, b, c) = ((ab + (a + b)c))c - abc + 0 = abc - abc = 0$$

Hence,
$$f(a, b, c) = k(a + b)(b + c)(c + a)$$

It is possible to determine the constant $k$ by expanding both sides and compare their corresponding coefficients. An alternative approach is to use the special value technique discussed in *Section 2.2* on *page 6*. Setting $a = b = c = 1$ gives

$$f(1, 1, 1) = 16 \quad \Longrightarrow \quad 16 = k \cdot 2 \cdot 2 \cdot 2 \implies k = 2$$

Therefore, we conclude

$$f(a, b, c) = \boxed{2(a + b)(b + c)(c + a)}$$

## 8.2.2 Symmetric Equations and Inequalities

Similar to polynomials, many equations and inequalities are symmetric too. In such cases, it may be possible to utilize symmetric based substitution to simplify the solution. Let's first review an example.

---

**Example 8.2.2**

Solve this system of equations:

$$\begin{cases} x + y &= 4 \\ xy &= 3 \end{cases}$$

---

This problem is not difficult at all. One can use regular substitution to cancel $y$ and obtain the following quadratic equation

$$x(4 - x) = 3 \implies x^2 - 4x + 3 = 0 \tag{8.1}$$

Alternatively, by Vieta's theorem, one can assert that $x$ and $y$ must be the two roots of the following equation:

$$t^2 - 4t + 3 = 0$$

Though obtained differently, these two quadratic equations are the same. In this particular example, either approach is sufficient. Meanwhile, this problem also presents a good example of utilizing a symmetric based substitution.

*Solution*

Let $x = 2 + t$ and $y = 2 - t$, where 2 is the average of $x$ and $y$ because $x + y = 4$.

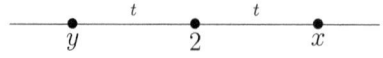

Then, we have

$$xy = 3 \implies (2+t)(2-t) = 3 \implies t^2 = 1 \implies t = \pm 1 \quad (8.2)$$

Hence, all the solutions are $(x, y) = \boxed{(3,1), (1,3)}$.

Done.

By comparison, the resulting quadratic equation *(8.2)* is symmetric and slightly simpler than *(8.1)* to solve. The difference in this particularly simple example is not significant. However, such symmetric based substitution is a powerful approach to solve some other problems. Let's review another example.

---

### Example 8.2.3

Let $a$ and $b$ be non-negative real numbers and $a + b = 2$. Show that:

$$\frac{1}{a^2 + 1} + \frac{1}{b^2 + 1} \le \frac{2}{ab + 1}$$

---

*Proof*

Let $a = 1 + t$ and $b = 1 - t$ where $-1 \le t \le 1$ because both $a$ and $b$ are non-negative. Then the given to-be-proved inequality is equivalent to

$$\frac{1}{(1+t)^2 + 1} + \frac{1}{(1-t)^2 + 1} \le \frac{2}{(1+t)(1-t) + 1} \quad (8.3)$$

Now, it just takes a few algebraic transformation steps to show that *(8.3)* indeed holds.

The left side can be simplified to

$$\frac{(t^2 - 2t + 2) + (t^2 + 2t + 2)}{(t^2 - 2t + 2)(t^2 + 2t + 2)} = \frac{2t^2 + 4}{(t^2 + 2)^2 - (2t)^2} = \frac{2(t^2 + 2)}{(t^4 + 4)}$$

Its right side can be simplified to

$$\frac{2}{2 - t^2}$$

Hence, *(8.3)* is equivalent to

$$\frac{2(t^2 + 2)}{(t^2 + 2)^2 - (2t)^2} \leq \frac{2}{2 - t^2}$$

$$\Leftrightarrow \qquad \frac{t^2 + 2}{t^4 + 4} \leq \frac{1}{2 - t^2}$$

$$\Leftrightarrow \quad (2 + t^2)(2 - t^2) \leq t^4 + 4$$

$$\Leftrightarrow \qquad 4 - t^4 \leq 4 + t^4$$

The last relation obviously holds. Therefore, we conclude that the original inequality holds.

*QED*

### 8.2.3 Symmetric Indeterminate Equation

An indeterminate equation is an equation which contains more than one variable. Solving indeterminate equations is an important technique to master because such problems appear frequently in math competitions at all levels from AMC8 to IMO.

There are several different types of indeterminate equations. Many of them have well-known solutions. The book *Indeterminate Equation* written by the same author is dedicated to this topic. In this section, we discuss how to solve one family of symmetric indeterminate equations.

**Example 8.2.4**

Find all pairs of positive integers $(x, y)$ satisfying

$$\frac{1}{x} + \frac{1}{y} = \frac{5}{6} \tag{8.4}$$

*Solution*

Obviously, *(8.4)* is symmetric. As such, let's first assume $x \leq y$. If all the solutions under this assumption can be found, then the complete set of solutions can be obtained by simply switching the values of $x$ and $y$.

$$x \leq y \implies \frac{1}{x} \geq \frac{1}{y} \implies \frac{1}{x} \geq \frac{1}{2} \times \frac{5}{6} = \frac{5}{12}$$

The second step above holds because, as the larger one of these two terms, $1/x$ must be at least the average of their sum. Now, we have obtained a single variable inequality whose positive integer solutions can be easily obtained.

$$\frac{1}{x} \geq \frac{5}{12} \implies 5x \leq 12 \implies x = 1, 2$$

When $x = 1$, no positive integer $y$ exists such that *(8.4)* can hold. When $x = 2$, $y = 3$ is one solution. Hence, under the assumption $x \leq y$, there exists only one solution $(x, y) = (2, 3)$. It follows that the original equation has two possible solutions:

$$(x, y) = \boxed{(2, 3), (3, 2)}$$

*Done.*

When there are three terms involved, this technique can still be used by asserting that that largest one must be equal to or greater than their average. After having solve the largest variable, the original three-variable equation can be transformed to a two-variable one which can be solved in a similar way. The following Romanian Olympiad problem is such an example.

$$\frac{1}{x} + \frac{1}{y} + \frac{1}{z} = \frac{3}{5}$$

Sometimes, a relation can be transformed to such basic patterns. The $23^{rd}$ problem in 2015 AMC12B provides an excellent example. The condition that the volume and the surface area of a rectangular box are numerically equal can be expressed as $abc = 2(ab + bc + ca)$. This is equivalent to

$$\frac{1}{2} = \frac{1}{a} + \frac{1}{b} + \frac{1}{c}$$

## 8.3   Solving Counting Problems

Utilizing symmetry to solve counting problems is discussed in the book *Counting* by the same author. Its core idea is to find symmetric sets of to-be-counted objects. Because of symmetry, the numbers of objects in each of these sets must be equal. Hence, it may be possible to count each set indirectly. Let's consider the following example.

---

**Example 8.3.1**

Joe goes to the cinema with his sister Mary and his parents. They are sitting next to each other in a row of four seats. If Joe wants to sit on the left of Mary, but not necessarily immediately left, how many different sitting plans are there?

---

*Solution*

For any sitting plan, it is always possible to just switch Joe's and Mary's seats so that one arrangement has Joe sit left to Mary and the other has Mary sit left to Joe. These two are mirroring to each other.

Therefore, all the possible sitting plans without any restriction can be divided into two symmetric sets. One contains all the arrangements when Joe sits on the left to Mary. The other contains all the arrangements when Mary sits on the right to Joe. The counts of these two sets must be equal.

By the basic counting principles, the total number of unrestrictive sitting plans is $4! = 24$. Hence, the number of sittings plans meeting the given restriction is half of that, i.e. $\boxed{12}$.

*Done.*

## 8.4   Practice

### Practice 1

Suppose the following system has one unique real number solution, find the value of $m$ and solve this system.

$$\begin{cases} x^2 + y^2 &= z \\ x + y + z &= m \end{cases}$$

(China)

### Practice 2

Compute

$$C_{2017}^1 + 2C_{2017}^2 + 3C_{2017}^3 + \cdots + 2016C_{2017}^{2016} + 2017C_{2017}^{2017}$$

**Practice 3**

Factorize $x^3 + y^3 + z^3 - 3xyz$.

**Practice 4**

In triangle $ABC$, let $\angle A = 30°$, $AB = 4$, and $AC = 3$. Points $M$ and $N$ locate on $AB$ and $AC$, respectively. Find the minimal value of $(CM + MN + NB)$.

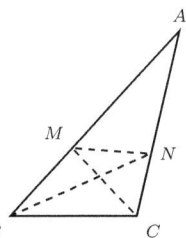

**Practice 5**

Ten balls, packed in a triangular crate, are either black or white, as shown. Prove there must exist three balls of the same color whose centers are vertices of an equilateral triangle.

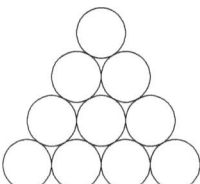

## Practice 6

Let $X$ be the integer part of $(3 + \sqrt{7})^n$ where $n$ is a positive integer. Show that $X$ must be odd.

## Practice 7

Let $n$ be a positive integer. Show that the smallest integer that is larger than $(1 + \sqrt{3})^{2n}$ is divisible by $2^{n+1}$.

## Practice 8

For pairwise distinct nonnegative real numbers $a, b, c$, prove that

$$\frac{a^2}{(b-c)^2} + \frac{b^2}{(c-a)^2} + \frac{c^2}{(b-a)^2} > 2$$

(2017 Canada MO)

## Practice 9

Let $n$ be an odd integer greater than 1. Let $\mathbb{A}$ be an $n \times n$ symmetric matrix such that each row and each column of $\mathbb{A}$ consists of some permutation of the integers $1, 2, \cdots, n$. Show that each one of the integers $1, 2, \cdots$ must appear in the main diagonal of $\mathbb{A}$.

## Practice 10

Find all positive integer solutions to this equation:

$$3(xy + yz + zx) = 4xyz$$

**Practice 11**

Find all integer solution to this equations:

$$\frac{1}{x} + \frac{1}{y} + \frac{1}{z} = \frac{3}{5}$$

(Romanian)

# Appendices

# Appendix A

# Solutions

## A.1   Introduction

This section is intentionally left blank.

So section numbers of solutions and practices can match.

## A.2 Go Simple and Go Extreme

### Practice 1

Brahmagupta's formula states that the area of a cyclic quadrilateral whose sides lengths are $a$, $b$, $c$, and $d$ is given by

$$S = \sqrt{(p-a)(p-b)(p-c)(p-d)}$$

where $p = \frac{1}{2} \cdot (a+b+c+d)$.

Describe its relationship with the Heron's formula.

Heron's formula states that the area of a triangle of side lengths $a$, $b$, and $c$ is given by

$$S = \sqrt{p(p-a)(p-b)(p-c)}$$

where $p = \frac{1}{2} \cdot (a+b+c)$.

It is a special case of Brahmagupta's formula. When two adjacent vertices of a quadrilateral coincide, this quadrilateral becomes a triangle. Accordingly, $d$ will equal 0 and Brahmagupta's formulas reduces to Heron's formula.

### Practice 2

It is known that the sum of the first $n$ cubes can be written as

$$1^3 + 2^3 + \cdots + n^3 = \left( \frac{1}{k} \cdot n \cdot (n+1) \right)^2$$

where $k$ is a positive constant. Determine the value of $k$.

Setting $n = 1$ yields

$$1^3 = \left( \frac{1}{k} \cdot 1 \cdot (1+1) \right)^2 \implies k = \boxed{2}$$

## Practice 3

Evaluate the value of $C_n^0 + 2C_n^1 + 4C_n^2 + \cdots + 2^n C_n^n$.

Setting $x = 2$ in binomial expansion *Equation 2.1* on *page 7*:

$$C_n^0 + 2C_n^1 + 4C_n^2 + \cdots + 2^n C_n^n = (1+2)^n = 3^n$$

## Practice 4

Find the sum of all the coefficients in the expanded form of

$$(x_1 + x_2 + \cdots + x_{2017})^{2017}$$

Assuming

$$(x_1 + x_2 + \cdots + x_{2017})^{2017} = a_1 x_1^{2017} + a_2 x_2^{2017} + \cdots$$

Each term on the right is a product of a coefficient (which can be 1) and some $x_i$ ($i = 1, 2, \cdots$). Then the desired sum equals $(a_1 + a_2 + \cdots)$. In order to find its value, let's set

$$x_1 = x_2 = \cdots = x_{2017} = 1$$

This will leads to

$$(1 + 1 + \cdots + 1)^{2017} = a_1 + a_2 + \cdots$$

Hence, the answer is $\boxed{2017^{2017}}$.

## Practice 5

Find the remainder when $x^{81} + x^{49} + x^{25} + x^9 + x$ is divided by $x^3 - x$.

Suppose

$$x^{81} + x^{49} + x^{25} + x^9 + x = (x^3 - x)Q(x) + R(x) \qquad \text{(A.1)}$$

where $R(x)$ is the remainder. Then the degree of $R(x)$ must be no more than 2 because the degree of $(x^3 - x)$ is 3. Let's assume

$$R(x) = ax^2 + bx + c \qquad \text{(A.2)}$$

The special value technique can be used to determine $R(x)$ without the need to perform conventional polynomial division. The key is to eliminate the unknown and actually irrelevant $Q(x)$. In order to achieve this, let's use the three roots of $(x^3 - x)$ as the special values. Setting $x = -1, 0, 1$, respectively, to *(A.1)* and also considering *(A.2)* lead to

$$\begin{cases} (-1)^{81} + (-1)^{49} + (-1)^{25} + (-1)^9 + (-1) &=& a(-1)^2 + b(-1) + c \\ (0)^{81} + (0)^{49} + (0)^{25} + (0)^9 + (0) &=& a(0)^2 + b(0) + c \\ (1)^{81} + (1)^{49} + (1)^{25} + (1)^9 + (1) &=& a(1)^2 + b(1) + c \end{cases}$$

Solving the above system yields $a = 0$, $b = 5$, and $c = 0$. Hence,

$$R(x) = \boxed{5x}$$

## Practice 6

Find all functions $f : \mathbb{R} \to \mathbb{R}$ such that

$$x^2 f(x) + f(1 - x) = 2x - x^4 \qquad \text{(A.3)}$$

Replacing $x$ with $(1 - x)$ yields

$$(1 - x)^2 f(1 - x) + f(x) = 2(1 - x) - (1 - x)^4 \qquad \text{(A.4)}$$

Solving *(A.3)*, *(A.4)* by treating $f(x)$ and $f(1 - x)$ as two variables gives

$$f(x) = \boxed{1 - x^2}$$

## Practice 7

Let the domain of function $f$ be all natural numbers and $f(1) = 1$. If for any natural numbers $m$ and $n$, it always hold that $f(m + n) = f(m) + f(n) + mn$, find $f(n)$.

Setting $m = 1$ and noting $f(1) = 1$ lead to

$$f(n + 1) = f(1) + f(n) + n = f(n) + (n + 1)$$

Setting $n = 1, 2, 3, \cdots, n$:

$$f(2) = f(1) + 2$$
$$f(3) = f(2) + 3$$
$$\cdots$$
$$f(n) = f(n - 1) + n$$

Adding these equations and canceling same terms yield

$$f(n) = f(1) + 2 + 3 + \cdots + n$$
$$= 1 + 2 + 3 + \cdots + n$$
$$= \boxed{\frac{n(n + 1)}{2}}$$

## Practice 8

A chocolate bar is made up of a rectangular $m$ by $n$ grid of small squares. Two players take turns to break up the bar. In each turn, a player picks a rectangular piece of chocolate and breaks it into two smaller ones by snapping along one whole line of subdivisions between its squares. The player who makes the last break wins. Does one of the players have a winning strategy for this game?

If the chocolate bar is $1 \times 2$, then obviously the $1^{st}$ player will win. If it is $1 \times 3$, then the $2^{nd}$ player will win. If it is $1 \times 4$, the $1^{st}$ player will win. We also examine some cases of $2 \times m$ and find, in such cases, the $1^{st}$ player always win.

Therefore, we conjuncture if the product of $m$ and $n$ is even, the $1^{st}$ player will win. Otherwise, if the product is odd, the $2^{nd}$ player will win.

To prove this conjuncture, we note that the game starts with 1 rectangular piece and ends with $(m \times n)$ pieces. In each turn, when a player breaks one piece into two, the total number of pieces increase by 1. Hence, the game will last $(mn - 1)$ turns. Obviously, when this value is odd which is equivalent to $mn$ is even, the $1^{st}$ player will win. Otherwise the $2^{nd}$ player will win.

### Practice 9

Suppose there are an infinite number of airports on a flat field which extends infinitely in all directions. No two of these airports are exactly the same distance apart. At a point in time, one plane will take off from each airport and land at its nearest neighboring airport. What is the maximum number of planes that may land at the same airport?

If there are just two airports, then the two planes will just swap their airports. This means the maximum number of planes landed at one airport will be one.

If there are three airports. To simplify, let's assume they are on the same straight line. Then the airport in the middle will receive two planes.

It is not difficult to show that it is possible for an airport to receive three planes at a time by simply adding an additional airport vertical to the middle airport.

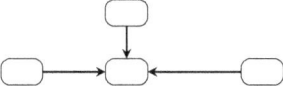

Clearly, it is also possible for an airport to receive four planes. Then, the question becomes whether it is possible to continuously add more airports so that planes departed from them will all arrive at this middle airport? It is not difficult to see that this is impossible. This is because when there are too many airports added, it will become too crowded which means that distances among these airports will be shorter than their distances to the intended receiving airport.

In order to quickly find the limit, let's relax the condition that requires their distances to be different. Instead, let's investigate the case when distances can be the same.

Assuming $A$ is the airport that receives the maximum number of planes. It is easy to see that this airport can receive up to 6 planes from six nearby airports located at the vertices of a regular hexagon centered at $A$.

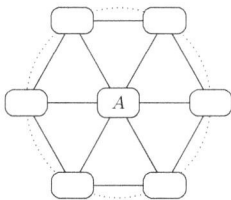

In this case, the distances among these airports are all the same. It is obvious that no more airport can be squeezed in without jeopardizing the assumption that $A$ is their nearest neighboring airport.

Because this is an extreme case when distances can be equal, the final answer to the original puzzle is $\boxed{5}$. These five airports can be first put on the outfit circle and then slightly moved so that their distances become different.

### Practice 10

There are 2017 people standing in a circle, numbered from 1 to 2017 sequentially. Starting from No. 1, they count alternating 1 and 2. People who count 2 will be out. The process continues until only one person left. What is his number?

Let's do several experiments and try to find some clues.

| No of people | The remaining person |
|:---:|:---:|
| 1 | 1 |
| 2 | 1 |
| 3 | 3 |
| 4 | 1 |
| 5 | 3 |
| 6 | 5 |
| 7 | 7 |
| 8 | 1 |
| ... | ... |

With some observation and thinking, it appears that if the number of initial people is a power of 2, such as 1, 2, 4, 8, and so on, the last remaining person's original position is 1.

This conjuncture indeed holds because when the number of people is even, then each round of elimination will get rid of half of people and No. 1 will still count 1 in the next round. If after each round of elimination, the number of remaining people is still even, then the process will continue and No. 1 will always count 1 until the last round. In order to ensure every round of elimination leaves an even number of survivors, the initial count must be a power of 2.

When the initial number is not a power of 2, we can wait till the remaining count becomes a power of 2. Whoever is due to count when the remaining number becomes a power of 2 will be the last survival.

For example, if there are 3 people initially, one person must be eliminated before the number of remaining people becomes a power of 2. The $1^{st}$ person to be eliminated is No. 2. At that time, it is No. 3's turn. Hence the last remaining person is 3.

When there are 5 people initially, one person must be out before the total count becomes 4 which is a power of 2. Hence the last remaining person is 3. When the initial number is 6, two people must be eliminated. They are No. 2 and No. 4, which makes the last remaining person originally stand at 5.

Using this logic, when the initial count is 2017, $2017 - 2^{10} = 993$ people must be eliminated before the headcount becomes a power of 2. Hence, the last remaining person must be $2 \times 993 + 1 = \boxed{1987}$.

## Practice 11

Let $p$ be an odd prime number. For positive integer $k$ satisfying $1 \leq k \leq (p-1)$, the number of divisors of $(kp+1)$ between $k$ and $p$, exclusive, is $a_k$. Find the value of $(a_1 + a_2 + \ldots + a_{p-1})$.

(2016 Japan MO)

Let's first try to find some clues by examining a few cases:

When $p = 3$:

| $p = 3$ | $k$ | $kp + 1$ | $a_i$ | divisors |
|---|---|---|---|---|
| | 1 | 4 | 1 | $(2)$ |
| | 2 | 7 | $-$ | $-$ |
| | | $a_1 + a_2 =$ | $\boxed{1}$ | $(2)$ |

When $p = 5$:

| $p = 5$ | $k$ | $kp + 1$ | $a_i$ | divisors |
|---|---|---|---|---|
| | 1 | 6 | 2 | $(2, 3)$ |
| | 2 | 11 | $-$ | $-$ |
| | 3 | 16 | 1 | $(4)$ |
| | 4 | 21 | $-$ | $-$ |
| | | $a_1 + \cdots + a_4$ | $\boxed{3}$ | $(2, 3, 4)$ |

When $p = 7$:

| $p = 7$ | $k$ | $kp + 1$ | $a_i$ | divisors |
|---|---|---|---|---|
| | 1 | 8 | 2 | $(2, 4)$ |
| | 2 | 15 | 2 | $(3, 5)$ |
| | 3 | 22 | $-$ | $-$ |
| | 4 | 29 | $-$ | $-$ |
| | 5 | 36 | 1 | $(6)$ |
| | 6 | 43 | $-$ | $-$ |
| | | $a_1 + \cdots + a_6$ | $\boxed{5}$ | $(2, 3, 4, 5, 6)$ |

It appears that

1) The answer is $\boxed{p - 2}$,

2) Every number in $2, 3, \cdots, p-1$ appears exactly once among the divisors, and

3) The largest number, i.e. when $k = p - 1$, does not have any qualified divisor.

The $3^{rd}$ point is obvious because a qualified divisor must be between $k$ and $p$ exclusive. By the same reasoning, any divisor $m$ can only possibly contribute to $a_1$, $a_2$, $\cdots$, $a_{m-1}$. Hence, the $2^{nd}$ observation above seems to be equivalent to asserting every integer $m$ $(1 < m < p)$ will contribute exactly once to $a_1$, $a_2$, $\cdots$, $a_{m-1}$. If this can be proved, then we can confidently assert that the answer is $\boxed{p-2}$.

This conjecture indeed holds. To prove it, let's first consider the following $(m-1)$ numbers.

$$p, \quad 2p, \quad 3p, \quad \cdots, \quad (m-1)p$$

Because $p$ is prime and $1 < m < p$, we find $m$ and $p$ must be relatively prime. It follows that the remainders of these $(m-1)$ numbers being divided by $m$ must be distinct. This is because if two of them have the same remainder, then their difference must be a multiple of $m$ which is impossible.[1] Meanwhile, none of these $(m-1)$ numbers is a multiple of $m$. Therefore, these $(m-1)$ remainders must be 1, 2, $\cdots$, $(m-1)$. Correspondingly, the remainders of the following $(m-1)$ numbers must be 0, 2, 3, $\cdots$, $(m-1)$:

$$p+1, \quad 2p+1, \quad 3p+1, \quad \cdots, \quad (m-1)p+1$$

This means that exactly one of them is a multiple of $m$.

### Practice 12

Find all functions $f : \mathbb{R} \to \mathbb{R}$ such that

$$f(yf(x) - x) = f(x)f(y) + 2x \qquad (A.5)$$

for all $x$, $y \in \mathbb{R}$.

(2016 Japan MO)

---

[1] Let these two numbers be $ip$ and $jp$ where $i > j$. Then their difference is $(i-j)p$. Because $(i-j) < m$ and $p$ is a prime greater than $m$, this difference cannot be a multiple of $m$.

Setting $x = y = 0$ leads to:

$$f(0) = f^2(0) \implies f(0) = 0, 1$$

We now discuss these two possibilities separately:

1) If $f(0) = 0$, then setting $y = 0$ leads to

$$f(-x) = 2x \implies \boxed{f(x) = -2x}$$

This is one solution.

2) If $f(0) = 1$, then setting $y = 0$ leads to

$$f(-x) = f(x) + 2x \qquad\qquad (A.6)$$

Replacing $x$ with $(yf(x) - x)$ in the $(A.6)$ leads to

$$
\begin{aligned}
f(x - yf(x)) &= f(yf(x) - x) + 2(yf(x) - x) \\
&= (f(x)f(y) + 2x) + 2(yf(x) - x) \qquad \because (A.5) \\
&= f(x)f(y) + 2yf(x) \\
&= f(x)(f(y) + 2y) \\
&= f(x)f(-y) \qquad\qquad\qquad \because (A.6)
\end{aligned}
$$

$$\therefore \qquad f(x - yf(x)) = f(x)f(-y)$$

Replacing $y$ with $-y$ and applying $(A.5)$ gives

$$f(x + yf(x)) = f(x)f(y)$$
$$\implies \quad f(x + yf(x)) = f(yf(x) - x) - 2x$$

Setting $y = -\frac{x}{f(x)}$ in the last relation yields

$$f(0) = f(-2x) - 2x \implies 1 = f(-2x) - 2x \implies f(-2x) = 1 - (-2x)$$

Finally, replacing $-2x$ with $x$ leads to another solution

$$f(x) = \boxed{1 - x}$$

Therefore, there exist totally two solutions

$$f(x) = -2x \qquad \text{and} \qquad f(x) = 1 - x$$

# A.3　Induction and Recursion

### Practice 1

Let $\{a_n\}$ be a sequence defined as $a_1 = 1$ and $a_n = \frac{a_{n-1}}{1+a_{n-1}}$ when $n \geq 2$. Find the general formula of $a_n$.

Let's compute a few terms by hand first aiming to find the general formula, and then apply the mathematical induction to prove the result.

$$a_1 = 1, \quad a_2 = \frac{1}{2}, \quad a_3 = \frac{1}{3}, \quad a_4 = \frac{1}{4}, \quad \cdots$$

Hence, a natural guess is that $a_n = \frac{1}{n}$. Now, let's prove it.

When $n = 1$, $a_1 = \frac{1}{1}$. Assuming $a_k = \frac{1}{k}$, then when $n = k+1$,

$$a_{k+1} = \frac{a_k}{1 + a_k} = \frac{\frac{1}{k}}{1 + \frac{1}{k}} = \frac{1}{k+1}$$

Therefore, we conclude that the general formula is $a_n = \frac{1}{n}$.

### Practice 2

An ATM machine can only dispense two-dollar bills and five-dollar bills. Show that it is always capable of dispensing exactly $n$ dollars when $n \geq 4$.

When $n = 4$, dispensing two 2-dollar bills will satisfy the requirement. Suppose it can dispense $k \geq 4$ dollars, we are going to show that it will be able to dispense $(k+1)$ dollars.

There are two possible scenarios for the composite of the first $k$ dollars: it contains at least one five-dollar bill and it contains no five-dollar bill.

1) If there is at least one five-dollar bill, then replacing one five-dollar bill with three two-dollar bills will make the total $(k+1)$ dollars.

2) If there is no five-dollar bill, then it must contain at least two two-dollar bills because $k \geq 4$. In this case, replacing these two two-bills with a five-dollar bill will make the total $(k+1)$ dollars.

Therefore, this ATM is always possible to dispense $(k+1)$ dollars which implies it is capable of dispensing $n$ dollars if $n \geq 4$.

### Practice 3

Show that for any positive integer $n$, it is always hold that

$$2^n + 2 > n^2$$

It is easy to show that this relation holds when $n = 1, 2, 3$.

Assuming when $n = k \geq 3$, it holds that $2^k + 2 > k^2$. Then when $n = k + 1$, we find

$$
\begin{aligned}
& 2^{k+1} + 2 \\
= & 2 \times 2^k + 2 \\
> & 2 \times (k^2 - 2) + 2 \qquad (\because 2^k + 2 > k^2 \implies 2^k > k^2 - 2) \\
= & 2k^2 - 2 \\
= & k^2 + (k^2 - 3) + 1 \\
\geq & k^2 + 2k + 1 \qquad (\because k \geq 3 \implies k^2 - 3 \geq 2k) \\
= & (k + 1)^2
\end{aligned}
$$

Thus, we conclude the relation holds for any natural number.

## Practice 4

Show that it is always possible to cut $n$ squares of arbitrary sizes into some pieces and then use these pieces to construct a single bigger square.

First, let's show that it is always possible to combine two squares into one square.

Suppose their side lengths are $a$ and $b$, respectively, and $a \geq b$. As shown in the diagram below, extend $BC$ to $J$ so that $CJ = b$ and take $K$ on $AB$ so that $AK = b$. Then $DKFJ$ is the desired combined bigger square because $\triangle DAK \cong \triangle DCJ$ and $\triangle KEF \cong \triangle JFG$.

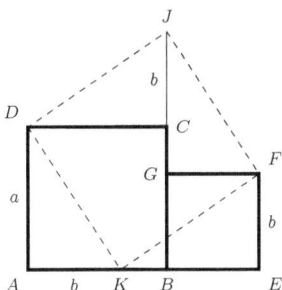

Now, assuming it is possible to combine $k \geq 2$ squares into one. Then for $(k+1)$ squares, we can first combine $k$ of them into one which results in two squares. As shown in the $n = 2$ case above, these two squares can be combined into one. Therefore, it is always possible to combine $(k+1)$ squares into one.

By the principle of mathematical induction, we conclude it is always possible to combine $n$ squares into one.

### Practice 5

Show that

$$1 \cdot 2^2 + 2 \cdot 3^2 + \cdots + n \cdot (n+1)^2 = \frac{n(n+1)}{12} \cdot (an^2 + bn + c)$$

holds for any positive integer $n$, where $a$, $b$, and $c$ are constants.

By *Example 2.2.1* on *page 6*, we find $a = 3$, $b = 11$, and $c = 10$. Here, we prove this relation holds for any $n$ using induction.

It is easy to show the relation holds when $n = 1$. Assume it also holds when $n = k$, i.e.

$$1 \cdot 2^2 + 2 \cdot 3^2 + \cdots + k \cdot (k+1)^2 = \frac{k(k+1)}{12} \cdot (3k^2 + 11k + 10)$$

then when $n = k + 1$:

$$1 \cdot 2^2 + 2 \cdot 3^2 + \cdots + k \cdot (k+1)^2 + (k+1)(k+2)^2$$

$$= \frac{k(k+1)}{12} \cdot (3k^2 + 11k + 10) + (k+1)(k+2)^2$$

$$= \frac{1}{12} \cdot k(k+1)(k+2)(3k+5) + (k+1)(k+2)^2$$

$$= \frac{1}{12} \cdot (k+1)(k+2)\Big(k(3k+5) + 12(k+2)\Big)$$

$$= \frac{1}{12} \cdot (k+1)(k+2)(3k^2 + 5k + 12k + 24)$$

$$= \frac{1}{12} \cdot (k+1)((k+1)+1)(3(k+1)^2 + 11(k+1) + 10)$$

Therefore, by the principle of mathematical induction, this relation holds for every positive integer $n$.

### Practice 6

Find a prime number $p$ so that it always divides $(3^{2n+1} + 2^{n+2})$ where $n$ is a positive integer.

Let $S_n = 3^{2n+1} + 2^{n+2}$. We are going to first make an educated guess of what this prime number may be. Afterwards, we can prove the conclusion using mathematical induction.

$$\begin{aligned} S_1 &= 3^{2 \times 1 + 1} + 2^{1+2} &= 35 &= 5 \times 7 \\ S_2 &= 3^{2 \times 2 + 1} + 2^{2+2} &= 259 &= 7 \times 37 \end{aligned}$$

Therefore, if there exists such a prime, it must be $\boxed{7}$. Assume $7 | S_k$, then when $n = k + 1$ we have

$$\begin{aligned} S_{k+1} &= 3^{2(k+1)+1} + 2^{(k+1)+2} \\ &= 9 \times 3^{2k+1} + 2 \times 2^{k+2} \\ &= 2 \times (3^{2k+1} + 2^{k+2}) + 7 \times 3^{2k+1} \\ &= 2 \times S_k + 7 \times 3^{2k+1} \end{aligned}$$

Because both terms are multiples of 7, its sum must be a multiple of 7, i.e., $7 | S_{k+1}$. By the principle of mathematical induction, it must be true that 7 always divides $(3^{2n+1} + 2^{n+2})$.

## Practice 7

Show that it always hold for any integer $n > 1$ that

$$\log_{10}(n!) > \frac{3n}{10} \cdot \left( \frac{1}{2} + \frac{1}{3} + \cdots + \frac{1}{n} \right)$$

When $n = 2$, we have

$$\log_{10}(2!) > \frac{3 \times 2}{10} \times \frac{1}{2} \quad \Leftrightarrow \quad 10 \log_{10} 2 > 3 \quad \Leftrightarrow \quad 2^{10} > 10^3 \qquad (A.7)$$

which does hold. Assume the relation holds when $n = k \geq 2$, i.e.,

$$\log_{10}(k!) > \frac{3k}{10} \cdot \left( \frac{1}{2} + \frac{1}{3} + \cdots + \frac{1}{k} \right) \qquad (A.8)$$

then when $n = k + 1$, by AM-GM inequality, we have

$$\frac{1 + \cdots + k}{k} \geq \sqrt[k]{1 \cdots k} \quad \Leftrightarrow \quad \frac{\frac{1}{2}k(k+1)}{k} \geq \sqrt[k]{k!} \quad \Leftrightarrow \quad k + 1 \geq 2 \cdot (k!)^{\frac{1}{k}}$$

Therefore,

$\log_{10}((k+1)!)$
$= \log_{10}(k!(k+1))$
$\geq \log_{10}(k! \cdot 2 \cdot (k!)^{\frac{1}{k}})$
$= \log_{10}(2 \cdot (k!)^{\frac{1}{k}+1})$
$= \log_{10}(2!) + \dfrac{k+1}{k}\log_{10}(k!)$
$> \dfrac{3 \cdot 2}{10} \cdot \dfrac{1}{2} + \dfrac{k+1}{k} \cdot \dfrac{3k}{10} \cdot \left(\dfrac{1}{2} + \dfrac{1}{3} + \cdots + \dfrac{1}{k}\right)$ $\quad (\because (A.7), (A.8))$
$= \dfrac{3}{10} + \dfrac{3(k+1)}{10} \cdot \left(\dfrac{1}{2} + \dfrac{1}{3} + \cdots + \dfrac{1}{k}\right)$
$= \dfrac{3(k+1)}{10} \cdot \left(\dfrac{1}{2} + \dfrac{1}{3} + \cdots + \dfrac{1}{k} + \dfrac{1}{k+1}\right)$

By the principle of mathematical induction, we conclude that the relation holds for any integer $n$ greater than 1.

### Practice 8

**(Tower of Hanoi)** Given a stack of $n$ disks of different sizes arranged in a neat stack in ascending order of size on one rod (the smallest at the top, thus making a conical shape), together with two empty rods, the towers of Hanoi puzzle asks for the minimum number of moves required to move the stack from one rod to another, where a move is only allowed if it places a smaller disk on a bigger one.

This is a classical puzzle which can be solved using recursion and induction. Clearly, only 1 move is required if $n = 1$. When $n = 2$, 3 moves will be required.

Let $S_n$ be the minimal number of moves required to accomplish a $n-$disk case. Then to move $(n + 1)$ disks, we need the following 3 steps:

i) Move the top $n$ disks to the $2^{nd}$ rod. This requires $S_n$ moves.

ii) Move the largest disk to the $3^{rd}$ rod. This requires 1 move.

iii) Move all the $n$ disks from the $2^{nd}$ to the $3rd$ rod. This requires $S_n$ moves.

All together, $(2S_n + 1)$ moves are quired to move $(n + 1)$ disks. Therefore, the recursion is

$$S_{n+1} = 2S_n + 1 \qquad \text{or} \qquad S_n = 2S_{n-1} + 1$$

Solving such recursion is discussed in the book *Competition Algebra*. Here is a solution using continuous iteration:

$$\begin{aligned}
S_n &= 2S_{n-1} + 1 \\
&= 2(2S_{n-2} + 1) + 1 \\
&= 2^2 S_{n-2} + (2 + 1) \\
&= 2^2(2S_{n-3} + 1) + (2 + 1) \\
&= 2^3 S_{n-3} + (2^2 + 2 + 1) \\
&= \cdots \\
&= 2^{n-1} S_1 + (2^{n-2} + \cdots + 2^2 + 2 + 1) \\
&= 2^{n-1} \times 1 + (2^{n-1} - 1) \\
&= \boxed{2^n - 1}
\end{aligned}$$

## Practice 9

Let $n$ be a positive integer. Show that

$$\left(1 + \frac{1}{3}\right)\left(1 + \frac{1}{3^2}\right)\cdots\left(1 + \frac{1}{3^n}\right) < 2$$

(China)

Let $f(n) = \left(1 + \frac{1}{3}\right)\left(1 + \frac{1}{3^2}\right)\cdots\left(1 + \frac{1}{3^n}\right)$.

Instead of showing $f(n) < 2$, let's prove a stronger conclusion $f(n) < 2 - \frac{1}{3^n}$ using mathematical induction.

When $n = 1$, $f(1) = 1 + \frac{1}{3} < 2 - \frac{1}{3}$. The claim holds.

Suppose the claim holds when $n = k$, i.e. $f(k) < 2 - \frac{1}{3^k}$. Then when $n = k + 1$,

$$f(k+1) = f(k)\left(1 + \frac{1}{3^{k+1}}\right)$$
$$< \left(2 - \frac{1}{3^k}\right)\left(1 + \frac{1}{3^{k+1}}\right)$$
$$= 2 - \frac{1}{3^k} + \frac{2}{3^{k+1}} - \frac{1}{3^{2k+1}}$$
$$< 2 - \frac{1}{3^k} + \frac{2}{3^{k+1}}$$
$$= 2 - \frac{1}{3^{k+1}}$$

Therefore, by the principle of mathematical induction, it always hold that
$$f(n) < 2 - \frac{1}{3^n} < 2$$

## Practice 10

Let $n$ be an integer greater than 2, prove $n^{n+1} > (n+1)^n$.

When $n = 3$, $3^{(3+1)} = 81 > 64 = 4^3$. The assertion holds.

Assuming the relation holds when $n = k > 2$, i.e.

$$k^{k+1} > (k+1)^k \tag{A.9}$$

Then, when $n = k + 1$, we only need to show

$$\frac{(k+1)^{k+2}}{k^{k+1}} > \frac{(k+2)^{k+1}}{(k+1)^k} \tag{A.10}$$

in order to prove $(k+1)^{k+2} > (k+2)^{k+1}$. This is because, by the assumption *(A.9)*, the denominator of *(A.10)*'s left side is greater than that of its right side. If *(A.10)* holds, then the numerator on its left side must be greater than that of its right side.

In order to show *(A.10)* hold, it is sufficient to show

$$
\begin{aligned}
& (k+1)^{k+2}(k+1)^k \;>\; (k+2)^{k+1}k^{k+1} \\
\Longleftrightarrow\quad & (k+1)^{2k+2} \;>\; (k+2)^{k+1}k^{k+1} \\
\Longleftrightarrow\quad & ((k+1)^2)^{k+1} \;>\; ((k+2)k)^{k+1} \\
\Longleftrightarrow\quad & (k+1)^2 \;>\; (k+2)k \\
\Longleftrightarrow\quad & k^2+2k+1 \;>\; k^2+2k
\end{aligned}
$$

The last relation clearly holds. Therefore, $(k+1)^{k+2} > (k+2)^{k+1}$ holds. It follows that the original assertion holds by the principle of mathematical induction.

### Practice 11

Find all functions $f : \mathbb{Q} \to \mathbb{Q}$ such that the Cauchy equation

$$f(x+y) = f(x) + f(y)$$

holds for all $x, y \in \mathbb{Q}$ where $\mathbb{Q}$ is the set of all rational numbers.

Setting $x = y = 0$ gives $f(0) = f(0) + f(0) \implies f(0) = 0$.

Now, let's prove $f(kx) = kf(x), k \in \mathbb{N}, x \in \mathbb{Q}$ using mathematical induction.

Obviously, when $k = 1$, it always hold that $f(x) = f(x)$. Assuming it is true for $k$, then for $k+1$:

$$f((k+1)x) = f(kx+x) = f(kx)+f(x) = kf(x)+f(x) = (k+1)f(x)$$

Thus, $f(kx) = kf(x)$ for all natural number $k$.

Now let's show $f(kx) = kf(x)$ holds for all integers $k$, regardless of positive or negative.

Setting $y = -x$:

$$f(0) = f(x) + f(-x) \implies 0 = f(x) + f(-x) \implies f(-x) = -f(x)$$

$$\therefore \qquad f(-kx) = -f(kx) = -kf(x)$$

This means, for any integer $k$, both positive and negative, $f(kx) = kf(x)$ always hold where $x$ is rational.

Now, taking $y = 1/k$ then

$$f(1) = f(k(1/k)) = kf(1/k) \implies f(1/k) = (1/k)f(1)$$

Finally, let $x$ be a rational number $\frac{m}{n}$ where both $m$ and $n$ are integers.

$$f(x) = f(m/n) = mf(1/n) = (m/n)f(1) = xf(1)$$

Hence, $f(x) = cx$ where $c = f(1)$ is a constant.

# A.4    Proof by Contradiction

### Practice 1

There are 13 squares of side length 1 positioned inside a circle of radius 2. Show that at least two of these squares have a common point.

If no two squares share a common point, then the total areas these squares cover is the sum of their individual areas, which is 13. However, the area of this circle is $4\pi \approx 12.56 < 13$. Hence, it is impossible.

### Practice 2

Prove that no integers $x$ and $y$ can satisfy $x^2 - 4y = 2$.

Let's assume there exist integers $x$ and $y$ satisfying $x^2 - 4y = 2$. Then $x^2$ must be even which implies $x$ is even. If $x$ is even, $x^2$ must be a multiple of 4. This means the left side $(x^2 - 4y)$ is a multiple of 4. However, the right side is not a multiple of 4. This is impossible which means no such integers exist.

### Practice 3

If there exist two integers $x$ and $y$ so that $ax + by = 1$ where both $a$ and $b$ are integers, show that $a$ and $b$ must be co-prime.

Suppose $a$ and $b$ are not co-prime, let their greatest common divisor be $d$ $(d > 1)$. It follows that the left side $(ax + by)$ must be a multiple of $d$. This will force the right side to be a multiple of $d$ too. However, this is clearly impossible because 1 cannot be divisible by $d > 1$. As a result, $a$ and $b$ must be co-prime.

### Practice 4

If all sides of a convex pentagon $ABCDE$ are equal in length and $\angle A \geq \angle B \geq \angle C \geq \angle D \geq \angle E$, show that $ABCDE$ is a regular pentagon.

It is sufficient to show $\angle A = \angle B = \angle C = \angle D = \angle E$. If this is not true, then it must have $\angle A > \angle E$.

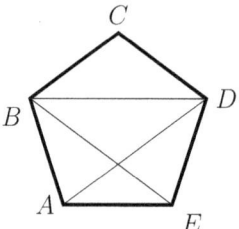

As shown, both $\triangle ABE$ and $\triangle EAD$ are isosceles because $AB = AE = DE$. Then the assumption $\angle A > \angle E$ will mean $BE > AD$. Applying this conclusion on $\triangle ABD$ and $\triangle EBD$ yields $\angle BDE > \angle DBA$. Next, because $BC = CD$, we have $\angle CDB = \angle CBD$. Combining these two means $\angle D > \angle B$. This is contracting to the given condition of $\angle B \geq \angle D$.

Therefore, it cannot be true that $\angle A > \angle E$ which means the assumption of $\angle A > \angle E$ is false. Thus, $\angle A = \angle E$ which in turn means all the angles are equal.

### Practice 5

Given $n > 2$ points on a plane. Prove if any straight line passing two of these points also passes another point, then all these $n$ points are collinear.

If these points are not collinear, then for any straight line passing any two two points, there exists at least one point which does not locate on this line. Let the distance from this point to that line be $h_i$, $i = 1, 2, \cdots$. Assume the smallest of these $h_i$'s be $h$ which is the distance between point $A$ and line $\ell$. Because $\ell$ passes at least three points, let them be point $B$, $C$, and $D$ in that order, as shown.

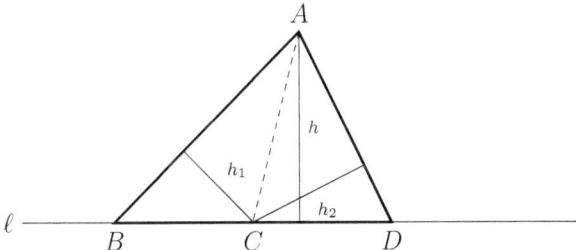

Let the distance from $C$ to $AB$ and $C$ to $AD$ be $h_1$ and $h_2$, respectively. Then we have $h_1 \geq h$ and $h_2 \geq h$. It follows that

$$
\begin{aligned}
S_{\triangle ABD} &= S_{\triangle ABC} + S_{\triangle ADC} \\
&= \frac{1}{2} \cdot AB \cdot h_1 + \frac{1}{2} \cdot AD \cdot h_2 \\
&\geq \frac{1}{2} \cdot AB \cdot h + \frac{1}{2} \cdot AD \cdot h \\
&= \frac{1}{2} \cdot (AB + AD) \cdot h \\
&> \frac{1}{2} \cdot BD \cdot h \\
&= S_{\triangle ABD} \\
\therefore \quad S_{\triangle ABD} &> S_{\triangle ABD}
\end{aligned}
$$

This is impossible. Therefore, the assumption made earlier cannot hold which means the original assertion is true.

## Practice 6

Let the lengths of five line segments be $a_1$, $a_2$, $a_3$, $a_4$, and $a_5$, respectively, where $a_1 \geq a_2 \geq a_3 \geq a_4 \geq a_5$. If any three of these five line segments can form a triangle, then prove at least one of such triangles is acute.

Assume that there is no acute triangle, then all triangles must be either right or obtuse. Consider the three triangles whose sides are $(a_1, a_2, a_3)$, $(a_2, a_3, a_4)$, $(a_3, a_4, a_5)$, then following relations must hold:
$$a_1^2 \geq a_2^2 + a_3^2, \quad a_2^2 \geq a_3^2 + a_4^2, \quad a_3^2 \geq a_4^2 + a_5^2$$

Adding these three gives $a_1^2 \geq a_3^2 + 2a_4^2 + a_5^2$. It follows that
$$a_1^2 \geq a_4^2 + 2a_4 a_5 + a_5^2 = (a_4 + a_5)^2 \implies a_1 \geq a_4 + a_5$$

The last relation means line segments with lengths $a_1$, $a_4$, and $a_5$ cannot form a triangle. This contradicts to the given condition which means the assumption of no acute triangle is false.

## Practice 7

Prove if $x$ satisfies $0 < x < \frac{\pi}{2}$, then $\sin x + \cos x > 1$.

Because $0 < x < \frac{\pi}{2}$, both $\sin x$ and $\cos x$ are positive. If the claim does not hold, then we must have
$$\begin{aligned}
& 0 < \sin x + \cos x \leq 1 \\
\Leftrightarrow \quad & 0 < (\sin x + \cos x)^2 \leq 1 \\
\Leftrightarrow \quad & 0 < \sin^2 x + 2\sin x \cos x + \cos^2 x \leq 1 \\
\Leftrightarrow \quad & 0 < 1 + 2\sin x \cos x \leq 1 \\
\Leftrightarrow \quad & -1 < 2\sin x \cos x \leq 0
\end{aligned}$$

Because neither $\sin x$ nor $\cos x$ equals 0, this means one of $\sin x$ and $\cos x$ is negative which contradicts the fact that both of them are positive. Hence, it must hold that $\sin x + \cos x > 1$.

## Practice 8

Let $a_0$, $a_1$, $\cdots$, $a_n$ be all integers. If $a_0$, $a_n$ and the sum of $(a_0 + a_1 + \cdots + a_n)$ are all odd integers, show that the following equation has no rational roots:

$$a_0 x^n + a_1 x^{n-1} + \cdots + a_{n-1} x + a_n = 0$$

Suppose this equation has a rational root $\frac{p}{q}$ where integer $p$ and $q$ are relatively prime. Then

$$a_0 \left(\frac{p}{q}\right)^n + a_1 \left(\frac{p}{q}\right)^{n-1} + \cdots + a_{n-1}\left(\frac{p}{q}\right) + a_n = 0$$

Multiplying both sides by $q^n$ yields

$$a_0 p^n + a_1 p^{n-1} q + \cdots + a_{n-1} p q^{n-1} + a_n q^n = 0 \qquad \text{(A.11)}$$

Therefore

$$a_0 p^n = -q(a_1 p^{n-1} + \cdots + a_{n-1} p q^{n-2} + a_n q^{n-1})$$

Because $p$ and $q$ are relatively prime, it must hold that $q$ divides $a_0$. This means $p$ is odd. Similarly, $p$ must divide $a_n$ because *(A.11)* can be rewritten as

$$a_n q^n = -p(a_0 p^{n-1} + a_1 p^{n-2} q + \cdots + a_{n-1} q^{n-1})$$

This means $q$ is odd too.

Now because both $p$ and $q$ are odd, then every term in *(A.11)*, $a_k p^{n-k} q^k$, has the same odd-even parity as $a_k$, where $k = 0, 1, \cdots, n$. It follows that the parity of the left side of *(A.11)* is the the same as the value of $(a_0 + a_1 + \cdots + a_{n-1} + a_n)$ which is odd. However, its right side, 0, is even. This is impossible. This contradiction means that the assumption made early cannot hold. It follows that the given equation has no rational root.

# A.5 Pigeonhole Principle

### Practice 1

Show that among any randomly placed 5 points inside an equilateral triangle whose side length is 2, the shortest distance between any two of these five points can not be longer than 1.

Divide this equilateral triangle into four smaller equilateral triangles of unit side length, as shown below.

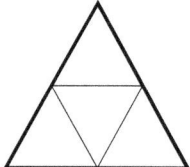

By the pigeonhole principle, at least two of these points will locate in the same smaller triangle. It is obvious that their distance cannot be longer than 1.

### Practice 2

Nine points are randomly placed in a unit square. Show that it is possible to select three of them to form a triangle whose area is no larger than $\frac{1}{8}$.

Let's divide this square into 4 equally spaced horizontal strips as shown below. Then each stripe is a $(1 \times \frac{1}{4})$ rectangle. By the pigeonhole principle, at least three points will be in one of these stripes. Then, the area of the triangle formed by these three points cannot exceed $\frac{1}{2} \times 1 \times \frac{1}{4} = \frac{1}{8}$.

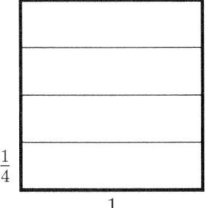

## Practice 3

There are only two problems in a math test. Ten points will be awarded for every correct answer. Two points will be given for any skipped problem. No point will be given for wrong answer. The teacher claims there must be at least 3 students who will receive a same score. Can you figure out the minimal number of students taking this test?

There are four possible total scores: 20, 12, 4, and 0. In order to confidently claim at least 3 students have the same score, there should be at least $4 \times 3 + 1 = \boxed{13}$ students attending this test.

## Practice 4

A box contains a large quantity of four different types of Easter eggs. One kid is allowed to take one or two eggs of his choice from this box. What is the minimal number of kids must be there in order to confidently assert that at least two kids make the same choice?

The answer is $\boxed{11}$. This is because there are 4 different choices for a kid to take just one egg and 6 different choices for taking 2 eggs. Hence, there exist totally 10 different possible choices.

## Practice 5

Given any five integers, show that it is always possible to select three of them so that their sum is a multiple of 3.

Let's just consider the remainders of these five integers being divided by 3. If the sum of these remainders is a multiple of 3, then the sum of the corresponding numbers must be a multiple of 3 too.

There are only three possible choices: 0, 1, and 2.

If at least three of them are the same, then their sum must be a multiple of 3.

Otherwise, if no three are the same, then at most two can be the same. Because there are totally five numbers, all these three possible remainders must present. Accordingly, their sum must be a multiple of 3.

Therefore, there must exist three remainders whose sum is a multiple of 3. Therefore, it is always possible to pick up three of these five numbers whose sum is a multiple of 3.

## Practice 6

Show that among any $(n + 1)$ integers, one can always find two of them whose difference is divisible by $n$.

This problem is an extension to *Example 5.1.1* on *page 37* and can be proved in a similar manner.

## Practice 7

Given 12 different 2-digit integers, show that one can always choose two of them so that their difference is a two-digit integer with identical unit and tens digits.

By the conclusion of the previous practice, we can always find two out of these 12 numbers such that their difference is a multiple of 11. Call this difference as $N$.

As the difference of two 2-digit numbers, $N$ can not have more than 2 digits. Meanwhile, because all these original 12 integers are distinct, $N$ cannot be zero. Hence, $N$ must be a two-digit integer. Meanwhile, because $N$ is a multiple of 11, its unit and tens digits must be the same.

## Practice 8

Show that there exists a multiple of 2017 whose digits are either 8 or 0.

Among the following 2018 integers:

$$8, 88, 888, \cdots, \underbrace{888\cdots8}_{2018}$$

it is always possible to find two of them whose difference is a multiple of 2017. By construction of these numbers, it is easy to see that this difference must be in the form of $88\cdots800\cdots0$.

## Practice 9

Show that any convex pentagon must have three vertices $A$, $B$, and $C$ such that $\angle ABC \leq 36°$.

The sum of a pentagon's five interior angles equals $540°$. Hence, the smallest of them cannot exceed $540 \div 5 = 108°$.

Meanwhile, this smallest interior angle can be further divided into three angles using the two edges and two diagonals from its vertex. The smallest among them cannot exceed $108 \div 3 = 36°$.

## Practice 10

Each point of a circle is colored either red or blue.

(a) Prove that there always exists an isosceles triangle inscribed in this circle such that all its vertices are colored the same.

(b) Does there always exist an equilateral triangle inscribed in this circle whose three vertices are colored the same?

(Philippines)

(a) Consider the five vertices of an inscribed regular pentagon. These five points must be on the circle. By the pigeonhole principle, at least three of them must be of the same color. It is easy to show that they must form an isosceles triangle.

Hints to solve this problem: the target is to find three points of same color. Because each point can have two different colors. Hence, by pigeonhole principle, we can at most utilize five points, i.e. among five points, there must be three having the same color. Then the question becomes what are the five special points we can choose on a circle?

(b) The answer is no. An counter example is a circle half of which is red and the other half is blue. In such a case, at least two vertices of an equilateral triangle must fall on the same semi-circle of one color. Hence, it is impossible to find a qualifed equilateral triangle in this case.

## Practice 11

A chess board is an $8 \times 8$ grid. A bishop can attack another piece on the same diagonal as it locates. What is the maximum number of bishops can be placed on a chess board peacefully with no one being able to attack another?

The answer is $\boxed{14}$. We are going to prove this by showing there are at most 14 such bishops using the pigeonhole principle and then presenting a position containing 14 bishops at peace.

First, we notice that a bishop in any black square and a bishop in any white square will never attack each other because they will never be on a same diagonal (see the diagram below, ignore the numbers). Hence, bishops in black squares and bishops in white squares form two isolated symmetric worlds. As a result, it is sufficient to just consider bishops in white squares.

There are totally seven diagonals in one direction, let's mark all the squares on one diagonal with one number to create seven disjoint sets, as shown. If two bishops are in the same set, they can attack each other.

If there are eight bishops on white squares, then at least two of them will be on the same diagonal by the pigeonhole principle. This implies that there cannot be more than 7 bishops on white squares in order to avoid mutual attack. Similarly, at most 7 bishops can be placed on black squares peacefully. Together, there can not be more than 14 bishops on a chess board.

Meanwhile, the following configuration shows it is indeed possible to have 14 bishops on a chess board with no one being able to attack another. Hence the answer is 14.

### Practice 12

Show that in a $n$-people party, at least two of them have met the same number of other guests before.

If there exits one person who has not met anyone else, then other guests can meet up to $(n-2)$ people before (excluding himself and the person who has not met anyone). Therefore, the number of people these guests have met before ranges from 0 to $(n-2)$. By the pigeonhole principle, at least two out of these $n$ people have met the same number of guests before.

If everyone has met at least one other guest in the party, then the number of people met before ranges from 1 to $(n-1)$. Hence, by the pigeonhole principle, at least two people must have met the same number of other guests before.

Either way, the claim holds.

## Practice 13

Two different sized roulettes share the same center. Each is divided into 200 equal sections. On the larger roulette, 100 randomly picked sections are colored in red and the rest are colored in green. The 200 sections on the smaller roulette are colored randomly in either red or green. The number of red sections and that of green section may be different. Show that it is always possible to rotate these two roulettes to a position so that at least 100 paired sections have the same color.

Pick up any section on the smaller roulette. When the smaller roulette finishes a complete rotation, this section will pair with each of the 200 sections on the larger roulette which creates 200 color combinations. Because there are equal numbers of red and green sections on the larger roulette, 100 of these color combinations will have the same color and the other 100 will have different colors.

Because there are 200 sections on the smaller roulette, totally $200 \times 100 = 20,000$ such color combinations will have the same color.

Meanwhile, there are 200 different relative positions between the two roulettes. Hence, by the pigeonhole principle, there must be one position satisfying the condition that at least $20,000 \div 200 = 100$ paired sections have the same color.

## Practice 14

Show that no matter how 10 people form a queue, there must exist at least four of them whose heights are in either ascending or descending order. These four people may or may not be next to each other.

Let's prove a generalized assertion here: given a sequence of any $(n^2 + 1)$ numbers, there must exist a subsequence containing $(n + 1)$

elements which are in either ascending or descending order. The original problem is equivalent to the case when $n = 3$.

We are going to show that if there exists no $(n + 1)$ elements which are in ascending order, then it must exist at least $(n + 1)$ elements which are in descending order.

Let these $(n^2 + 1)$ numbers be $\{a_1, a_2, \cdots, a_{n^2+1}\}$. For any given $a_i$, let $m_i$ be the length of the longest sub-sequence in ascending order. By the assumption that there exists no $(n + 1)$ elements in ascending order, we have $1 \leq m_i \leq n$ holds for all $m_i$, $i = 1, 2, \cdots, n^2 + 1$.

Therefore, we have $(n^2+1)$ numbers taking values from $\{1, 2, \cdots, n\}$. By the pigeonhole principle variation 2, there must exist at least $(n + 1)$ equal ones. Let them be $m_{i_1} = m_{i_2} = \cdots = m_{i_{n+1}}$. Accordingly, the first element of these subsequences are $a_{i_1}, a_{i_2}, \cdots, a_{i_{n+1}}$, respectively.

If $a_{i_1} \leq a_{i_2}$, then the combination of $a_{i_1}$ and the $m_{i+1}$ numbers starting with $a_{i_2}$ in ascending order will create a new sequence in ascending order whose length is $m_{i_2} + 1$. This implies $m_{i_1} > m_{i_2}$ which is a contradiction. Therefore, $a_{i_1} > a_{i_2}$ must hold.

By similar reasoning, $a_{i_2} > a_{i_3} > \cdots > a_{i_{n+1}}$ must hold too. However, this situation will mean $a_{i_1}, a_{i_2}, \cdots, a_{i_{n+1}}$ form a sequence in descending order.

## Practice 15

Show that every integer $k > 1$ has a multiple which is less than $k^4$ and can be written in base 10 using at most 4 different digits.

(IMO Short List Problem)

Choose $n$ such that $2^{n-1} \leq k < 2^n$. Let $\mathbb{S}$ be the set of non-negative integers less than $10^n$ that can be written with digits 0 or 1 only. Clearly, $\mathbb{S}$ has $2^n$ elements.

Because $k < 2^n$, therefore it is always possible to find two elements $a > b$ in $\mathbb{S}$ whose difference $(a - b)$ is a multiple of $k$.

Notices that both $a$ and $b$ contain only digits 0 and 1. By studying the possible cases of subtracting one such number by another such number with possible carries, we can conclude $(a - b)$ can be expressed with digits 0, 1, 8, 9 only.

Meanwhile, the largest number in $\mathbb{S}$ whose digits are either 1 or 0 is $\underbrace{11\cdots1}_{n-1}$ and the smallest one is 0. Therefore,

$$a - b \leq \underbrace{11\cdots1}_{n-1} - 0 < 10^{n-1} \times 1.2 < 16^{n-1} = (2^{n-1})^4 \leq k^4$$

This proves that the number $(a - b)$ meets all the requirements.

# A.6   The Coloring Method

### Practice 1

There are 24 lily pads on a pond as shown below. Frank, the frog, wants to visit all these pads without stopping at the same pad more than once. He can jump to a neighboring pad either horizontally or vertically, but not diagonally. If Frank can choose any pad as its starting point, can he achieve his goal?

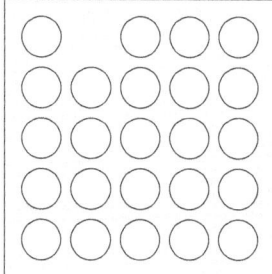

The answer is no. To show this, let's color these 24 lily pads in alternating white and black colors as shown.

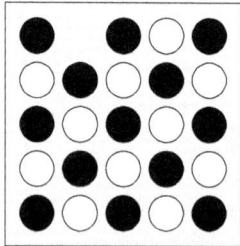

This gives a total of 13 black lily pads and 11 white lily pads. By the rules, Frank must jump to a lily pad of a different color in each move. Therefore, 24 moves can only visit 12 white and 12 black pads. This creates a contradiction.

**Practice 2**

In the center unit of a $3 \times 3 \times 3$ cubic lives a bug. Two units which share a face are connected via a door. The bug wants to visit all the units starting from his own without repeating. Is it possible?

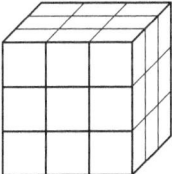

The answer is no. To prove this conclusion, let's color the 27 units in alternating black and white, as shown. There will be 14 black units and 13 white ones. The center unit will be white.

Because two neighboring units have different colors, the bug must visit units in alternating colors starting from a white one. Hence, it cannot visit 14 black units and 13 white ones in 27 moves. This means its wish cannot be fulfilled.

### Practice 3

Show that if an $m \times n$ grid can be completely covered by some L-shaped grids consist of 4 unit grids without overlapping, then the product of $m$ and $n$ must be a multiple of 8.

(Beijing)

Clearly, because each piece has 4 grids, the product of $m$ and $n$ must be a multiple of 4. To show that the product is a multiple of 8, it is sufficient to show that an even number of such L-shaped pieces are required. This conclusion is a natural extension to *Example 6.1.2* on *page 49* which has shown an odd number of such pieces cannot cover a rectangular board.

### Practice 4

Show that it is impossible to cover an $8 \times 8$ board using fifteen $4 \times 1$ pieces and one $2 \times 2$ piece.

Let's color the board in the following way:

Each $4 \times 1$ piece covers equal number of black and white squares. But a $2 \times 2$ will cover 3 squares of 1 color and 1 square of the other color. Hence, fifteen $4 \times 1$ pieces and one $2 \times 2$ piece cannot cover equal number of black and white squares. However, this board has exactly same number of squares in each color. Hence, it is impossible to cover this board using these pieces.

### Practice 5

Is it possible to cover a $6 \times 6$ grid using one L-shaped piece made of 3 grids and eleven $3 \times 1$ smaller grids?

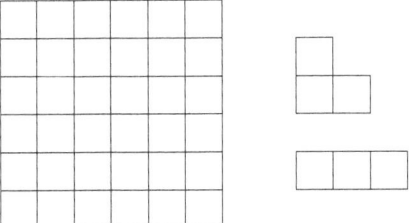

It is impossible. To show this, let's color the $6 \times 6$ grid using three different colors in the following way.

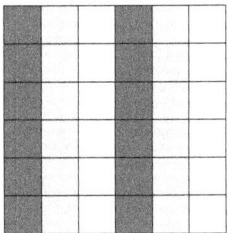

It is clear that each $1 \times 3$ piece will cover three grids either of the same color (call it a same-colored piece) or of three different colors (call it a multi-colored piece). The L-shaped piece will cover two grids of the same color and one with another color. Without loss of generality, let's assume the L-shaped piece cover 2 grids of color one and one grid of color two. Furthermore, let's assume the number of same-colored pieces which covers color 1, color 2, and color 3, are $x$,

$y$, and $z$, respectively. Finally, let's assume there are $d$ multi-color pieces. Hence, the total number of grids covered are:

color 1:   $3x + d + 2$

color 2:   $3y + d + 1$

color 3:   $3z + d$

These three numbers must be all equal according to our coloring scheme. This implies

$$3x + d + 2 = 3z + d \implies 3(x - z) = 2$$

However the last relation cannot hold because its left side is a multiple of 3, but the right side is not.

### Practice 6

It is possible to use some $1 \times 2 \times 4$ blocks to construct a $6 \times 6 \times 6$ cubic?

No, it is not possible. To prove this conclusion, let's color a $6 \times 6 \times 6$ cubic in the following way.

Then, there are a total of $8 \times 14 = 112$ dark unit blocks and $8 \times 13 = 104$ white ones. Each $1 \times 2 \times 4$ block will cover an equal number of dark unit blocks and white ones. Hence, the total number of dark unit blocks must equal to that of the white ones in order for a cubic to be constructed. Therefore, we conclude this is an impossible task.

### Practice 7

There are 6 points in a 3-D space. No three points are on the same line and no four points are one the same plane. Hence totally 15 segments can be created among these points. Show that if each of these 15 segments is colored either black or white, there must exist a triangle whose sides are of same color.

This problem is essentially the same as *Example 6.1.3* on *page 50*. Consequently, it can be solved in the same way.

### Practice 8

Seventeen people correspond by mail with one another - each one with all the rest. In their letters only three different topics are discussed. Each pair of correspondents deals with only one of these topics. Prove that there are at least three people who write to each other about the same topic.

(1964 IMO)

This problem is an extension of the previous practice. Seventeen people can be represented using 17 dots. The three topics can be presented using three different colors. The objective is to show that there must exist a triangle whose three edges are of the same color.

Pick an arbitrary point $A$. Among the 16 connections $A$ has with the remaining dots, at least 6 of them must be of the same color by the pigeonhole principle. Without loss of generality, let's name this color as red. If any two of these 6 are connected by a red edge, then we are done.. Otherwise, if none of them are connected by a red edge, then these 6 dots are connected each other in two different colors. By the conclusion of *Example 6.1.3* on *page 50*, there must exist a triangle whose edges are of the same color.

### Practice 9

Randomly color all the points in a plane using either white or black. Show that

1. There must exist a unit length segment whose two ends are colored same.

2. There must exist a right triangle whose three vertices are colored same.

Let's construct an equilateral triangle $ABC$ where $AB = BC = CA = 1$. By the pigeonhole principle, among its three vertices, at least two must have the same color. Then this edge satisfies the $1^{st}$ claim.

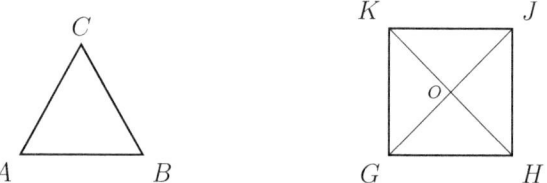

By the previous conclusion, there must exist a unit length segment whose ending points have the same color. Let such a segment be $GH$. Then draw a square $GHJK$ using $GH$ as one of its sides. If one of $J$ and $K$ has the same color as $GH$, then we find such a right triangle. If neither of them has the same color as $GH$ does, then $JK$ must have the same color themselves. Now examine the color of $O$, the intersection point of the two diagonals $GJ$ and $KH$. It is easy to see regardless of $O$'s color, either $\triangle OGH$ or $\triangle OJK$ must satisfy the $2^{nd}$ claim.

## Practice 10

Is it possible to arrange the numbers $1, 1, 2, 2, 3, 3, \cdots, 1986,$ 1986 into a sequence so that there is 1 number between two 1's, 2 numbers between two 2's, $\cdots$, 1986 numbers between two 1986's?

(China)

Let's color the $(1986 \times 2)$ positions in alternating black and white. Per given conditions (there are 1 number between two 1's, 2 numbers between two 2's, and so on), there are even number of intervals between two even numbers. So, a pair of even numbers must occupy one black and one white positions. Similarly, a pair of odd numbers must be separated by an odd number of spaces, thus they must occupy two positions of the same color.

Now, there are 993 pair of even numbers, therefore they will occupy 993 black and 993 white positions. Let $B$ and $W$ denote the number of paired odd numbers that occupy black and white positions, respectively. Then, it must hold that $B + W = 993$. Because each pair contains two numbers, the number of black positions and white positions occupied by these odd numbers will be $2B$ and $2W$, respectively.

Now, the total number of black positions equal $(993 + 2B)$ and the total number of white positions is $(993 + 2W)$. Clearly, they must be equal which means

$$993 + 2B = 993 + 2W$$

However, because $B + W = 993$ is odd, it must hold that $B \neq W$. Hence the above equality will never hold. It follows that the answer to the original question is no.

# A.7 Two-State Problem

### Practice 1

Find all pairs of prime numbers $(x, y)$ so that $x^2 + y = 2003$.

Because 2003 is odd, $x^2$ and $y$ must have different parities. Because there is only one even prime which is 2, hence one of $x$ and $y$ must be 2. When $x = 2$, we find $y = 1999$ which is prime. When $y = 2$, $x$ is not an integer. Therefore, this equation has only one prime solution

$$(x, y) = \boxed{(2, 1999)}$$

### Practice 2

Four $x$'s and five $o$'s are written around the circle in an arbitrary order. If two consecutive symbols are the same, then a new $x$ is inserted in between. Otherwise, a new $o$ is inserted. After nine new symbols are inserted, the previous 9 old ones are erased. Is it possible to get nine $o$'s after having repeated this operation for a finite time?

The answer is no. Let $x = 1$ and $o = -1$, then two consecutive symbols will be replaced by their product. Let $P$ and $P'$ be the product of the nine values before and after each operation, it is easy to see that $P' = P^2$ which should be always 1. This means after each operation, the product of these numbers must equal 1. However, nine $o$'s implies $P' = -1$. Therefore, it is impossible.

The above presented proof is a "standard" way to solve such types of problems mathematically. For this particular problem, there is a quick solution. In order to obtain nine $o$'s, it is necessary to have nine symbols arranged in an alternating way. However, it is impossible to have an alternating pattern with nine symbols.

### Practice 3

There are three piles of stones, numbering 19, 8, and 9, respectively. You are allowed to choose two piles and transfer one stone from each of them to the third pile. Is it possible to make all piles all contain exactly 12 stones after several such operations?

The answer is no. To see this, let's consider the remainder of stone count being divided by 3. Initially, these three remainders are 0, 1, and 2, respectively.

It is easy to see that after each move, these remainders will still be 0, 1, and 2. However, the desired ending scenario requires their remainders to be all 0. Therefore, it is impossible.

Hint to solve this problem: it is likely to be an odd-even problem. However, simply attempting odd-even (i.e. divide 2) appears to ineffective. Given there are three piles, how about trying dividing by 3?

### Practice 4

Prove: if $a$, $b$, $c$ are odd integers, then the quadratic equation $ax^2 + bx + c = 0$ has no integer solution.

Let's show that regardless of parity of $x$, $(ax^2 + bx + c)$ will be always odd, hence cannot equal to 0.

If $x$ is even, then both $ax^2$ and $bx$ are even. Because $c$ is odd, therefore $(ax^2 + bx + c)$ is odd. If $x$ is odd, all of $ax^2$, $bx$ and $c$ are odd because $a$, $b$, and $c$ are all odd. Therefore, $(ax^2 + bx + c)$ is always odd. This means it cannot equal 0.

## Practice 5

Let $f(x)$ be a polynomial with integer coefficients. If $f(1) = 3$, is it possible that $f(3) = 0$?

It is impossible. Assuming

$$f(x) = a_n x^n + a_{n-1} x^{n-1} + \cdots + a_1 x + a_0$$

then

$$f(1) = a_n + a_{n-1} + \cdots + a_1 + a_0$$

and

$$f(3) = a_n \cdot 3^n + a_{n-1} \cdot 3^{n-1} + \cdots + a_1 \cdot 3 + a_0$$

Because $a_k$ and $a_k \cdot 3^k$ $(k = 0, 1, \cdots, n)$ have same parity, their sum must have same parity too. This means that $f(1)$ and $f(3)$ will have the same parity. Therefore, if $f(1) = 3$, then $f(3)$ cannot equal 0.

## Practice 6

If the sum of $n$ integers equals 0 and their product equals $n$, show that $n$ must be a multiple of 4.

Let the $n$ integers be $a_1, a_2, \cdots, a_n$. Then

$$a_1 + a_2 + \cdots + a_n = 0 \tag{A.12}$$

$$a_1 \cdot a_2 \cdots \cdots a_n = n \tag{A.13}$$

If $n$ is odd, then by *(A.13)*, all $a_i$ are odd. It follows that the left side of *(A.12)* is the sum of an odd number of odd integers which must be odd. Therefore, the sum cannot equal 0.

Hence, $n$ must be even. This means that the product of these $n$ integers are even. This is equivalent to saying at least one of these

$a_i$ is even. Meanwhile, by *(A.12)*, there must be an even number of odd $a_i$ because their sum is even. It follows that at least two of $a_i$ are even. Now back to *(A.13)*, if at least two of them are even, their product must be a multiple of 4. Hence, $n$ must be a multiple of 4.

### Practice 7

Given any nine distinct points in 3-dimensional space with integer coordinates, show that there must exist two points so that the segment between them contains another point of integer coordinates.

Let the coordinates of these nine points be $(x_i, y_i, z_i)$ where $i = 1, 2, \cdots, 9$. Now, consider their midpoints $(\frac{x_i+x_j}{2}, \frac{y_i+y_j}{2}, \frac{z_i+z_j}{2})$, where $i, j = 1, 2, \cdots, 9$. If we can show that there exist a pair of $(x_i, y_i, z_i)$ and $(x_j, y_j, z_j)$ whose corresponding $(x_i, x_j)$, $(y_i, y_j)$ and $(z_i, z_j)$ all have the same parities, then the coordinates of their midpoint must be integer.

This can be proved using the Pigeonhole principle. Because there are $2^3 = 8$ choices for parities of an integer triplet $(x_i, y_i, z_i)$, there must exist two pairs having the same parities among 9 points.

### Practice 8

There are four types of cards costing 10, 15, 25 and 40 cents each. Joe bought totally 30 cards using several dollar bills exactly. If he bought 5 each of two types, and 10 each of the other two types, how much total did he spend on this purchase?

Let $a$, $b$, $c$, $d$ be the numbers of each type of cards, and $k$ be the number of dollar bills Joe spent. Then

$$
\begin{aligned}
10a + 15b + 25c + 40d &= 100k \\
\implies \qquad 2a + 3b + 5c + 8d &= 20k \\
\implies \qquad (2a + 8d) + (3b + 5c) &= 20k
\end{aligned}
$$

Because both $(2a+8d)$ and $20k$ are even, $(3b+5c)$ must be even too. This will lead to the conclusion that $3b$ and $5c$ have the same parity which implies $b$ and $c$ have the same parity.

If both of them are 10, then $a = d = 5$ which implies $k = 13/2$. This contradicts to the given condition that $k$ is an integer.

If both of them are 5, then $a = d = 10$ which implies $k = 7$. Therefore, Joe spent $\boxed{7}$ dollars on this purchase.

### Practice 9

There are $n$ points, $A_1$, $A_2$, $\cdots$, $A_n$ on a line segment, $\overline{A_0 A_{n+1}}$. The point $A_0$ is black, $A_{n+1}$ is white, and the rest points are colored randomly either black or white. Show that, among these $n+1$ line segments $A_k A_{k+1}$, where $k = 0, 1, \cdots, n$, the number of segments with different colored ending points is odd.

Mark a white point as $(1)$ and a black point as $(-1)$. For each line segment, we multiply the values of its two ending points. If the result is $(1)$, then the two ending points must have the same color. If the product is $(-1)$, then the two ending points' colors are different.

Because there are $(n+1)$ segments, there are $(n+1)$ products. If we multiply all these $(n+1)$ products, the result must be $(-1)$ because $A_1$, $A_2$, $\cdots$, $A_n$ all appear twice and $A_0 = -1$, $A_{n+1} = 1$ each appears once. This implies that, among the $(n-1)$ products, the number of $(-1)$ must be odd. This means that there are odd number of segments whose two ending points have different colors.

## Practice 10

Given two grids shown below, is it possible to transform (a) to (b) after a series of operations? In each operation, one can change all the signs in either one entire row or one entire column.

(Russia)

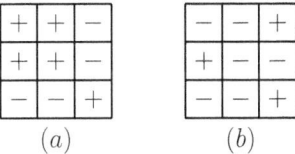

$(a)$ $(b)$

The answer is no. To show this, let's consider the top left $2 \times 2$ corner.

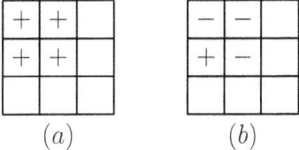

$(a)$ $(b)$

In (a), the number of '+' signs is even initially. After each operation, it will always remain as even. However, the number of '+' signs in (b) is odd. Hence, it is impossible to transform this $2 \times 2$ corner which implies it is impossible to transform the entire grid from (a) to (b).

## Practice 11

Let $0.a_1 a_2 a_3 a_4 \cdots$ be a decimal where $a_1$ is an odd integer and $a_2$ is an even integer. For $k > 2$, $a_k$ equals the unit digit of the value $(a_{k-1} + a_{k-2})$. Prove such a decimal must be rational.

In order to show a decimal is rational, it is sufficient to show that it has a repetend.

By the given definition, a digit is uniquely determined by its two preceding digits. If there exist two consecutive digits which have appeared before, then all the following digits will repeat.

Because $a_1$ is odd and $a_2$ is even, it is clear that the parities of $0.a_1a_2a_3a_4\cdots$ must be

$$0.OEOOEOOEOOEO\cdots$$

where $E$ indicates an even digit and $O$ indicates an odd digit. Obviously, there are infinite number of $OE$ pairs in this decimal:

$$0.O\underline{EO}OE\underline{OO}EO\underline{OE}O\cdots$$

However, there are only 5 distinct odd digits, $(1, 3, 5, 7, 9)$, and 5 distinct even digits, $(0, 2, 4, 6, 8)$. Therefore, there are totally 25 different possible $OE$ pairs. It follows that among the first 26 $OE$ pairs, at least two of them must be the same. Hence, such a decimal must be recurring.

## Practice 12

Katie had a collection of red, green and blue beads. She noticed that the number of beads of each color was a prime number and that the numbers were all different. She also observed that if she multiplied the number of red beads by the total number of red and green beads she obtained a number exactly 120 greater than the number of blue beads. How many beads of each color did she have?

(Scottish)

Let $r, g, b$ be the numbers of red, green and blue beads. Then we have

$$r(r + g) = 120 + b \tag{A.14}$$

Clearly, these three numbers cannot be all odd. Otherwise, if they are all odd, then the left side of *(A.14)* is even, but the right side is odd. This creates a contradiction.

Because $r, g$ and $b$ are distinct prime numbers, one of them must be 2.

If $b = 2$, then $r(r + g) = 122$. Factorizing 122 leads to $122 = 2 \times 62$ which will force $r = 2$ as well. This contracts the condition that these three numbers are distinct.

If $r = 2$, then the left side of *(A.14)* is even. Then $b$ has to be 2 in order for the right side to be even. This cannot hold either.

If $g = 2$, then $r(r + 2) = 120 + b$. Rearranging this relation as

$$b = r^2 + 2r - 120 \implies b = (r + 12)(r - 10)$$

Because $b$ is prime, it must hold that $r - 10 = 1$ which yield the only solution to the given problem

$$(r, g, b) = (11, 2, 23)$$

## Practice 13

An executioner lines up 100 prisoners single file and puts a red or a blue hat on each prisoner's head. Every prisoner can see the hats of the people in front of him in the line - but not his own hat, nor those of anyone behind him. The executioner starts at the end (back) and asks the last prisoner the color of his hat. He must answer "red" or "blue." If he answers correctly, he is allowed to live. If he gives the wrong answer, he is killed instantly and silently. (While everyone hears the answer, no one knows whether an answer was right.) On the night before the line-up, the prisoners confer on strategy to help them. What should they do in order to save as many prisoners as possible?

For the convince of discussion, let's call the prisoner standing at the back of the line as #1 because he is the first person to speak. Similarly, the prisoner standing in front of #1 is called #2, and so on.

Let's start investigating this problem by considering some simple cases.

If there is only one prisoner, there is no way to guarantee his survival because there is no any information available at this moment. In fact, regardless of the number of prisoners, it is always true that the fate of #1 cannot be guaranteed. Whatever #1 says is the first piece of information available to all other prisoners.

If there are two prisoners, while there is no guaranteed way to save himself, #1 can save #2 by calling out the color of #2's hat.

When there are three prisoners, clearly there is a guaranteed way to save at least one life. The question is whether there is a guaranteed way to save both #2 and #3? If so, this will be the best case.

Indeed, it is possible to save all the prisoners except #1 in a guaranteed way. One strategy to save both #2 and #3 is that #1 calls out

- red if he see the number of red hats in front of him is odd .

- blue if that number is even.

With this information, #2 can figure out the color of his hat by comparing the color of #3's hat. Similarly, #3 can also figure out the color of his hat be combining the information provided by #1 and #2.

It is easy to verify that this strategy works for any number of prisoners. Therefore, it is possible to save everyone except #1.

NOTES: The essence of this strategy is modular arithmetic operation which is discussed in the book *Number Theory - Modular Arithmetic* by the same author. In fact, a similar strategy exists to save all but the last prisoner even if hats have more than two colors.

# A.8 Symmetry

### Practice 1

Suppose the following system has one unique real number solution, find the value of $m$ and solve this system.

$$\begin{cases} x^2 + y^2 &= z \\ x + y + z &= m \end{cases}$$

(China)

The given system is symmetric with respect to $x$ and $y$. Therefore, if $(x, y, z)$ is one solution, so will be $(y, x, z)$. Because, it has only one unique solution, it must hold that $x = y$. Setting $y = x$ yields:

$$\begin{cases} 2x^2 &= z \\ 2x + z &= m \end{cases}$$

Canceling $z$ leads to a quadratic equation $2x^2 + 2x - m = 0$. As it has only one solution, we must have

$$2^2 + 4 \times 2 \times m = 0 \implies m = \boxed{-\frac{1}{2}}$$

Setting $m$ back can derive the solution to the original system as

$$(x, y, z) = \boxed{-\frac{1}{2}, -\frac{1}{2}, \frac{1}{2}}$$

### Practice 2

Compute

$$C_{2017}^1 + 2C_{2017}^2 + 3C_{2017}^3 + \cdots + 2016C_{2017}^{2016} + 2017C_{2017}^{2017}$$

Let

$$S = C_{2017}^1 + 2C_{2017}^2 + \cdots + 2016C_{2017}^{2016} + 2017C_{2017}^{2017}$$

Let's add a term of $0C_{2017}^0$ to $S$. (So a symmetric expression is constructed.)

$$S = 0C_{2017}^0 + 1C_{2017}^1 + \cdots + 2016C_{2017}^{2016} + 2017C_{2017}^{2017} \qquad (A.15)$$

Reversing the order of *(A.15)* and noting that $C_{2017}^{2017} = C_{2017}^0$, $C_{2017}^{2016} = C_{2017}^1$, and so on, lead to

$$\begin{aligned} S &= 2017C_{2017}^{2017} + 2016C_{2017}^{2016} + \cdots + 1C_{2017}^1 + 0C_{2017}^0 \\ &= 2017C_{2017}^0 + 2016C_{2017}^1 + \cdots + 1C_{2017}^{2016} + 0C_{2017}^{2017} \qquad (A.16) \end{aligned}$$

Adding *(A.15)* to *(A.16)* and consolidating same terms:

$$\begin{aligned} 2S &= 2017C_{2017}^0 + 2017C_{2017}^1 + \cdots + 2017C_{2017}^{2016} + 2017C_{2017}^{2017} \\ &= 2017(C_{2017}^0 + C_{2017}^1 + \cdots + C_{2017}^{2016} + C_{2017}^{2017}) \\ &= 2017 \times 2^{2017} \end{aligned}$$

Therefore we conclude $S = \boxed{2017 \times 2^{2016}}$.

Hints: the following well-known identity is proved in *Example 2.2.2* on *page 7*:

$$C_n^0 + C_n^1 + \cdots + C_n^n = 2^n$$

## Practice 3

Factorize $x^3 + y^3 + z^3 - 3xyz$.

Let $f(x, y, z) = x^3 + y^3 + z^3 - 3xyz$. By guessing and checking, it can be show that if $x + y + z = 0$, then $f(x, y, z) = 0$. This implies $(x + y + z)$ divides $f(x, y, z)$. Because $f(x, y, z)$ is a $3^{rd}$ degree symmetric polynomial, $(x + y + z)$ is one degree, the other factor must be a symmetric $2^{nd}$ degree polynomial. Therefore, it must hold that

$$f(x, y, z) = (x + y + z)[m(x^2 + y^2 + z^2) + n(xy + yz + zx)]$$

where $m$ and $n$ are to-be-determined coefficients.

Letting $x = y = z = 1$ yields $m + n = 0$.

Letting $x = y = 1$ and $z = -1$ yields $3m - n = 4$.

Solving the two equations above gives $m = 1$ and $n = -1$. Hence

$$f(x, y, z) = \boxed{(x + y + z)(x^2 + y^2 + z^2 - xy - yz - zx)}$$

## Practice 4

In triangle $ABC$, let $\angle A = 30°$, $AB = 4$, and $AC = 3$. Points $M$ and $N$ locate on $AB$ and $AC$, respectively. Find the minimal value of $(CM + MN + NB)$.

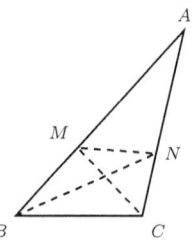

It is well known that the shortest path between two points is the straight segment. Hence, if we can transform this zig-zag path to one between two points, then the problem can be solved. This transformation can be done using reflection.

Let $B'$ be the mirroring point of $B$ with respect to $AC$, and $C'$ be the mirroring point of $C$ with respect to $AB$.

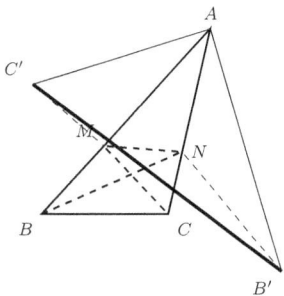

By symmetry, $BN = B'N$ and $CM = C'M$. Therefore

$$CM + MN + NB = C'M + MN + NB'$$

Clearly, the shortest path between $B'$ and $C'$ is the straight segment $B'C'$. Accordingly, the optimum locations of $M$ and $N$ are the intersection points of $B'C'$ and $AB$, $AC$, respectively.

Meanwhile, by symmetry again, $\angle BAC' = \angle BAC = \angle CAB'$. Therefore, $\angle C'AB' = 3 \times \angle BAC = 90°$. Applying Pythagorean theorem on $\triangle AB'C'$ gives

$$B'C' = \sqrt{AC'^2 + AB'^2} = \sqrt{AC^2 + AB^2} = \boxed{5}$$

### Practice 5

Ten balls, packed in a triangular crate, are either black or white, as shown. Prove there must exist three balls of the same color whose centers are vertices of an equilateral triangle.

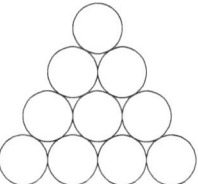

If this claim does not hold, let's assume the center ball is black without loss of generality. Then, let's consider the colors of the six balls which are adjacent to the center ball. Please refer to the figure 1 below.

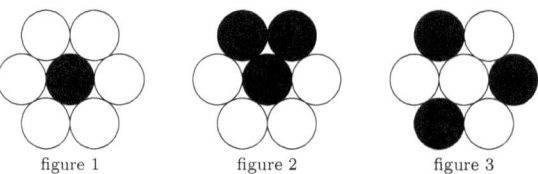

figure 1        figure 2        figure 3

If there are at least three black balls, then there must exist three black balls whose centers form an equilateral triangle. This is because if at least two of these three balls are adjacent, then a black triangle can be formed together with the center ball (see figure 2 above). Otherwise, if these three balls are not adjacent to each other, then they form a black triangle themselves (see figure 3 above).

Meanwhile, there cannot be five white balls. Because, otherwise it is always possible to pick up three of them to form a white triangle (similar to figure 3 above).

Therefore, in order for these six balls, together with the center black ball, not for form a same color triangle, they must be 2 black and 4 whites. Additionally, the two black balls must be opposite to each other because otherwise there will be either a black equilateral triangle or a white one. Without loss of generality, let's assume their positions are as shown in figure 4 below.

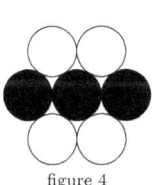

figure 4

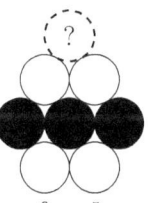

figure 5

Now, let's consider the original top ball in figure 5 above. If it is white, then a white triangle can be formed by this top ball and the two white balls on the $2^{nd}$ row. If it is black, then a black triangle can be formed by this top ball and the two side black balls on the $3^{rd}$ row.

Therefore, we conclude it is always possible to find three balls of the same color whose centers are vertices of an equilateral triangle.

## Practice 6

Let $X$ be the integer part of $(3 + \sqrt{7})^n$ where $n$ is a positive integer. Show that $X$ must be odd.

The symmetric counterpart of $(3+\sqrt{7})^n$ is $(3-\sqrt{7})^n$. Let's first consider the following number, instead:

$$Y = (3 + \sqrt{7})^n + (3 - \sqrt{7})^n \qquad (A.17)$$

It can be shown by binomial expansion that $Y$ is an even integer. This is because all the terms containing odd powers of $\sqrt{7}$ are canceled and the remaining terms with even powers of $\sqrt{7}$ are paired.

Next, we have $0 < 3 - \sqrt{7} < 1$ while means $(3 - \sqrt{7})^n$ is a positive number less than one too. Based on *(A.17)*, we have $Y$ is an even integer, therefore the integer part of $(3 + \sqrt{7})^n$ must equal $(Y - 1)$ which is odd.

### Practice 7

Let $n$ be a positive integer. Show that the smallest integer that is larger than $(1 + \sqrt{3})^{2n}$ is divisible by $2^{n+1}$.

By the same technique used in the previous practice, we find $(1 + \sqrt{3})^{2n} + (1 - \sqrt{3})^{2n}$ is an even integer. Let it be $2k$. Because $0 < |1 - \sqrt{3}| < 1$, the smallest integer that exceeds $(1 + \sqrt{3})^{2n}$ must be $2k$.

Hence, the problem is equivalent to show that

$$2^{n+1} \mid 2k = (1 + \sqrt{3})^{2n} + (1 - \sqrt{3})^{2n}$$

Let's compute the value of $2k$ using binomial expansion.

$$\begin{aligned}
2k &= (1 + \sqrt{3})^{2n} + (1 - \sqrt{3})^{2n} \\
&= (\sqrt{3} + 1)^{2n} + (\sqrt{3} - 1)^{2n} \\
&= ((\sqrt{3} + 1)^2)^n + ((1 - \sqrt{3} - 1)^2)^n \\
&= (4 + 2\sqrt{3})^n + (4 - 2\sqrt{3})^n \\
&= 2^n((2 + \sqrt{3})^n + (2 - \sqrt{3})^n)
\end{aligned}$$

By using the binomial expansion again, we find the term

$$(2 + \sqrt{3})^n + (2 - \sqrt{3})^n$$

is an even integer. Hence, $2^n((2 + \sqrt{3})^n + (2 - \sqrt{3})^n)$ must be a multiple of $2^{n+1}$.

### Practice 8

For pairwise distinct nonnegative real numbers $a, b, c$, prove that

$$\frac{a^2}{(b-c)^2} + \frac{b^2}{(c-a)^2} + \frac{c^2}{(b-a)^2} > 2$$

(2017 Canada MO)

Without loss of generality, let's assume $a$ is the smallest among these three numbers and set $b = a + x, c = a + y$. Then, both $x$ and $y$ are non-negative.

It follows that

$$
\begin{aligned}
& \frac{a^2}{(b-c)^2} + \frac{b^2}{(c-a)^2} + \frac{c^2}{(a-b)^2} \\
= {} & \frac{a^2}{(x-y)^2} + \frac{(a+x)^2}{y^2} + \frac{(a+y)^2}{x^2} \\
\geq {} & \frac{x^2}{y^2} + \frac{y^2}{x^2} \\
> {} & 2
\end{aligned}
$$

### Practice 9

Let $n$ be an odd integer greater than 1. Let $\mathbb{A}$ be an $n \times n$ symmetric matrix such that each row and each column of $\mathbb{A}$ consists of some permutation of the integers $1, 2, \cdots, n$. Show that each one of the integers $1, 2, \cdots$ must appear in the main diagonal of $\mathbb{A}$.

141

Because each of these integers appears in every row, therefore it must appear exactly $n$ times. Meanwhile, because the matrix is symmetric, therefore the number of any integer $k$, $k = 1, 2, 3, \cdots, n$, must appear the same times above the main diagonal as below the diagonal. Hence, it must appear not on the main diagonal even number of times. This implies that every integer must appear on the main diagonal at least once. However, there are only $n$ positions on the main diagonal to hold these $n$ integers. Thus, each one of these integers must appear once and only once on the main diagonal.

**Practice 10**

Find all positive integer solutions to this equation:

$$3(xy + yz + zx) = 4xyz$$

Because $x$, $y$ and $z$ are all non-zero, we can divide both sides by $xyz$. The result can be rearranged as

$$\frac{1}{x} + \frac{1}{y} + \frac{1}{z} = \frac{4}{3}$$

Assuming $x \leq y \leq z$, then

$$\frac{1}{x} \geq \frac{1}{y} \geq \frac{1}{z} \implies \frac{1}{x} \geq \frac{1}{3} \times \frac{4}{3} = \frac{4}{9} \implies x = 1, 2$$

When $x = 1$, we have

$$\frac{1}{y} + \frac{1}{z} = \frac{4}{3} - \frac{1}{1} = \frac{1}{3} \implies \frac{1}{y} \geq \frac{1}{2} \times \frac{1}{3} = \frac{1}{6}$$

Setting $y = 1$ to 6, respectively, finds $y = 4$, $z = 12$ and $y = 6$, $z = 6$ are the only two integer solutions under the assumption $x \leq y \leq z$.

When $x = 2$, we have

$$\frac{1}{y} + \frac{1}{z} = \frac{4}{3} - \frac{1}{2} = \frac{5}{6} \implies \frac{1}{y} \geq \frac{1}{2} \times \frac{5}{6} = \frac{5}{12}$$

Because $y \geq x = 2$, the only candidate is $y = 2$ here which leads to $z = 3$.

Here, we conclude all the solutions are all the permutations of $(1, 4, 12)$, $(1, 6, 6)$, and $(2, 2, 3)$.

### Practice 11

Find all integer solution to this equations:

$$\frac{1}{x} + \frac{1}{y} + \frac{1}{z} = \frac{3}{5}$$

(Romanian)

Using the same technique as demonstrated in the previous practice can find all the solutions to this Romain Olympiad problem are all the permutations of the following sets:

$(2, 11, 110)$, $(2, 12, 60)$, $(2, 14, 35)$, $(2, 15, 30)$, $(2, 20, 20)$, $(3, 4, 60)$, $(3, 5, 15)$, $(3, 6, 10)$, $(4, 4, 10)$, and $(5, 5, 5)$.